Mining the Biomedical Literature

Computational Molecular Biology
Sorin Istrail, Pavel Pevzner, and Michael Waterman, editors

Computational molecular biology is a new discipline, bringing together computational, statistical, experimental, and technological methods, which is energizing and dramatically accelerating the discovery of new technologies and tools for molecular biology. The MIT Press Series on Computational Molecular Biology is intended to provide a unique and effective venue for the rapid publication of monographs, textbooks, edited collections, reference works, and lecture notes of the highest quality.

Computational Molecular Biology: An Algorithmic Approach
Pavel A. Pevzner, 2000

Computational Methods for Modeling Biochemical Networks
James M. Bower and Hamid Bolouri, editors, 2001

Current Topics in Computational Molecular Biology
Tao Jiang, Ying Xu, and Michael Q. Zhang, editors, 2002

Gene Regulation and Metabolism: Postgenomic Computation Approaches
Julio Collado-Vides, editor, 2002

Microarrays for an Integrative Genomics
Isaac S. Kohane, Alvin Kho, and Atul J. Butte, 2002

Kernel Methods in Computational Biology
Bernhard Schölkopf, Koji Tsuda, and Jean-Philippe Vert, editors, 2004

An Introduction to Bioinformatics Algorithms
Neil C. Jones and Pavel A. Pevzner, 2004

Immunological Bioinformatics
Ole Lund, Morten Nielsen, Claus Lundegaard, Can Keşmir, and Søren Brunak, 2005

Ontologies for Bioinformatics
Kenneth Baclawski and Tianhua Niu, 2005

Biological Modeling and Simulation: A Survey of Practical Models, Algorithms, and Numerical Methods
Russell Schwartz, 2008

Combinatorics for Genome Rearrangements
Guillaume Fertin, Anthony Labarre, Irena Rusu, Éric Tannier, and Stéphane Vialettte, 2009

Learning in Computational and Systems Biology
Neil D. Lawrence, Mark Girolami, Magnus Rattray, and Guido Sanguinetti, 2010

Algorithms in Structural Molecular Biology
Bruce R. Donald, 2011

Mining the Biomedical Literature
Hagit Shatkay and Mark Craven, 2012

Mining the Biomedical Literature

Hagit Shatkay and Mark Craven

The MIT Press
Cambridge, Massachusetts
London, England

MIT Press books may be purchased at special quantity discounts for business or sales promotional use. For information, please email special_sales@mitpress.mit.edu or write to Special Sales Department, The MIT Press, 55 Hayward Street, Cambridge, MA 02142.

This book was set in Syntax and Times Roman by Toppan Best-set Premedia Limited. Printed and bound in the United States of America.

Library of Congress Cataloging-in-Publication Data

Shatkay, Hagit.
Mining the biomedical literature / Hagit Shatkay and Mark Craven.
   p.   cm.—(Computational molecular biology)
Includes bibliographical references and index.
ISBN 978-0-262-01769-5 (hardcover : alk. paper)
1. Medical literature—Data processing. 2. Biological literature—Data processing. 3. Data mining. 4. Medical informatics. 5. Bioinformatics. 6. Information storage and retrieval systems—Medicine. 7. Information storage and retrieval systems—Biology. 8. Content analysis (Communication). 9. Information retrieval. I. Craven, Mark. II. Title.
R118.6.S53   2012
610.285—dc23
2011047751

10 9 8 7 6 5 4 3 2 1

To Yaron, Eadoh, and Ruth
—HS

To Susan, Owen, and Kelly
—MC

Contents

Acknowledgments   ix

1   Introduction   1

2   Fundamental Concepts in Biomedical Text Analysis   9

3   Information Retrieval   33

4   Information Extraction   53

5   Evaluation   77

6   Putting It All Together: Current Applications and Future Directions   99

References   115

Index   131

Acknowledgments

As Selim Akl kindly pointed out at the very early stages of this endeavor, writing a book can be a pretty lonely job. However, as it turned out, our interaction with students and colleagues throughout this time—conversations, discussions, debates, and arguments, both public and private—all helped to shape and sharpen the ideas presented in this book. We would like to thank the many people who have contributed to our understanding and to our view of all that is discussed in this book, as well as those whose friendship and support made it possible for us to actually write it.

Many thanks go to our colleagues and collaborators in the biomedical text mining excursion, especially: Lynette Hirschman, Christian Blaschke, Alfonso Valencia, Russ Altman, Sophia Ananiadou, Lan Aronson, Christopher Baker, Sabine Bergler, Ted Briscoe, Bob Carpenter, Monica Chagoyen, Kevin Cohen, Aaron Cohen, Marc Colosimo, Dina Demner-Fushman, Anna Divoli, Ian Donaldson, Dave Eichmann, Noemie Elhadad, Carol Friedman, Udo Hahn, Marti Hearst, Bill Hersh, Eivind Hovig, Larry Hunter, Lars Jensen, Su Jian, Nikiforos Karamanis, Ian Lewin, Marc Light, Anália Lourenço, David McClosky, Yves Moreau, Alex Morgan, Sean O'donoghue, Alberto Pascual-Montago, Hoifung Poon, Sampo Pyysalo, Dietrich Rebholz-Schuhmann, Phoebe Roberts, Luis Rocha, Patrick Ruch, Andrey Rzhetsky, Padmini Srinivasan, Jun'ichi Tsujii, Karin Verspoor, Anne-Lise Veuthey, Anita de Waard, Mark Weeber, Bonnie Webber, Cathy Wu, Alex Yeh, Hong Yu, Ralf Zimmer, and the many others who have been part of the vivid discussions at TREC Genomics, BioCreative, BioLINK, and BioNLP workshops.

From all the above, we are particularly thankful to Lan Aronson for the information he provided about MeSH indexing, to Sean O'donoghue for the image of the Reflect system, to Eivind Hovig for the image of PubGene, as well as to Kevin Cohen who was a firm-enough believer in the book so as to start referencing it quite a few years before its completion.

Special thanks are due to John Wilbur for introducing HS to biomedical text in the first place and to Mark Boguski and Stephen Edwards for being her first biological partners venturing into this area. Colleagues and collaborators from several fields, with whom our interactions have always been stimulating and have often given rise to new research directions, are: Paul Ahlquist, Vineet Bafna, Asa Ben-Hur, Dorothea Blostein, Chris Bradfield, William Cohen, Art Delcher, John Denu, Dan Fasulo, Mike Flanigan, Iddo Friedberg, Lew Friedland, Ronen Feldman, Michael Gould, Annette Höglund, Jeanette Holden, Nathalie Japkowicz, Oliver Kohlbacher, Johan Kumlien, Stan Matwin, Jason Miller, Clark Mobarry, Parvin Mousavi, Robert Murphy, Sharon Regan, Hershel Safer, Ron Stewart, Ellen Voorhees, Shibu Yooseph, Malik Yousef, and Jerry Zhu. We thank all of them, and do apologize for unintentional omission of any names.

Both of us were fortunate enough to have several people who acted as mentors throughout the years, providing knowledge, insight, guidance and advice on science and on writing, along with encouragement and opportunities to pursue the directions leading to this book. For HS, these include Catriel Be'eri, Leslie Pack Kaelbling, Tom Dean, Gene Myers, Sorin Istrail and Granger Sutton. For MC, they include Jude Shavlik, Tom Mitchell and Dave DeMets. We are sincerely grateful to all of them.

HS also thanks both her former department chairs at Queen's University, Jim Cordy and Selim Akl, for supporting her in taking on the book writing, as well as Errol Lloyd, her department chair at the University of Delaware, for his consideration, which allowed her to complete it.

Many of our excellent students and postdocs explored some of the fundamental ideas presented in this book and helped demonstrate their value. Others helped us in exploring and learning about related ideas. HS thanks Scott Brady, Fengxia Pan, Rob Denroche, Phil Hyoun Lee, Andrew Wong, Jess Shen, Nawei Chen, Yin Lam, Quazi Abidur Rahman, Zhi Hong Zheng, Na Harrington, Limin Zheng, and Sara Salehi. MC thanks Dave Andrzejewski, Joe Bockhorst, Debbie Chasman, Deborah Muganda, Keith Noto, Irene Ong, Yue Pan, Soumya Ray, Burr Settles, Marios Skounakis, Adam Smith, and Andreas Vlachos.

We are grateful to our editors at MIT Press—Bob Prior, for keeping his high spirits and belief in this project over several long years, Susan Buckley and Katherine Almeida for shepherding the book through the production process—and to Sorin Istrail, the series editor, who invited us to write the book with MIT Press in the first place.

In terms of the biological aspects of this book, and specifically for enhancing our understanding about the work of database curators, there are several people to whom we are indebted: Rosanne Charlap-Orbach was the first to introduce HS, shortly after the sequencing of the fly, to FlyBase curation, complex pathways, and how publications and text relate to both. Other people who are well versed in the workings of the various organism databases and who were so readily willing to engage in discussions about their work and challenges are Jane Abu-Threideh, Evelyn Camon, Rolf Apweiler, Cecilia Arighi, Ramana Madupu, Martin Ringwald, Mary Shimoyama, and Judith Blake.

We also thank the several government agencies whose funding has supported our research in biomedical text mining. HS acknowledges Canada's NSERC Discovery Grant 298292–2004, 2009, NSERC Discovery Acceleration Supplement award 380478–2009, and Ontario's Early Researcher Award ER07–04–085, and MC acknowledges grant R01 LM07050 from the NIH National Library of Medicine and a National Science Foundation CAREER Award.

HS thanks several wonderful people (and her good fortune to have them close by) who were helpful throughout the process, encouraging, believing, inspiring, and otherwise making the world beautiful—and writing possible: Noga Ami-rav, Dee Fujita Kratzke, Nathalie Valette-Silver, Nancy Beller-Simms, Bonnie Eisenberg, Jill and Steve Gould, Robin Markowitz, Mollie Miedzinski, Sue Petito, Shulamit Oren, Naomi Orgil, and Michal Shatkay along with Rachel, Ma'ayan, and Ira Blevis. She is particularly grateful to the three who have borne the brunt of this project but were always there to offer every possible comfort and loving support: Yaron Reshef, Eadoh Reshef, and Ruth Shatkay.

MC would like to acknowledge the love, support, and patience of his family: Susan Goral, Owen Craven, and Kelly Craven. They have always provided a welcome and joyful respite from activities such as book writing.

1 Introduction

The current millennium started with the sequencing of the human genome. There are now thousands of sequenced genomes available, covering a wide range of organisms and a broad collection of individuals within the human population. Additionally, there is a multitude of datasets characterizing dynamic aspects of cells such as molecular abundances, interactions, and localizations. The hope is that in knowing and analyzing the sequences of such genomes and associated data, scientists are opening the "book of life" and will be able to understand the intricate processes governing life, death, and disease at the most basic molecular level.

However, the enterprise of understanding this book of life is one of enormous complexity, requiring the sustained efforts of many researchers working in a wide range of scientific areas. Knowledge about biological entities and processes has been acquired by thousands of scientists through decades of experimentation and analysis. This knowledge is often represented in text form. Much of it is published in the vast biomedical literature but there are many other sources of scientific information in text form, such as lab notebooks and web pages. Currently, there are numerous organized efforts focused on representing some of this information in structured and accessible formats within publicly available databases and repositories. The goal of such efforts is to enable scientists to quickly relate and compare new findings with previous ones in the hope of expediting productive discovery and research.

A notable characteristic of the current era is massive-scale production of data and information. For example, numerous types of high-throughput methods, including genome-scale sequencing, have transformed biology into a data-rich science. At the same time, there has been a steady and overwhelming increase in the number of scientific publications. For example, citations for almost four million articles were added to the

PubMed database [186] in the period from 2006 to 2010. This is more than twice as many as were added during a similar five-year period 20 years earlier. These parallel trends give rise to a situation in which biomedical researchers have more data than ever to analyze and interpret, while there is far more available background knowledge that must be taken into account for the analysis and interpretation of these data. Consider a typical case in which a high-throughput biological experiment results in significant responses for hundreds of genes. The scientific community's knowledge about these genes and their relationships to one another is distributed across numerous databases and thousands of articles. This situation, and many variations of it, calls for a significantly larger role to be played by automated text-analysis methods in the biomedical sciences.

Consequently, there has been a surge of interest in biomedical text analysis over the past decade. Researchers from a wide range of disparate communities—including natural language processing, information retrieval, biomedical informatics, and the life sciences—have contributed ideas and applications to this enterprise.

Despite the progress that has been made in biomedical text mining, this technology is not nearly as widely exploited in the biomedical domain as it could be. The goal of this book is to introduce researchers from a variety of backgrounds to the key ideas in biomedical text mining. In particular, we discuss (1) the distinct text-mining tasks that have been framed, (2) the principal challenges that are involved in addressing these tasks, (3) a broad and versatile toolbox of methods for accomplishing these tasks, (4) methodology for empirically evaluating text-mining systems, and (5) the ways in which text-mining methods are being applied to address several challenging problems in biomedicine.

This book provides a structured introduction to biomedical text mining from two perspectives. One perspective is application oriented. The main chapters of the book are organized around the principal tasks that are addressed by text-mining systems, and chapter 6 describes how such systems have been assembled and applied in several significant biomedical applications. A second perspective is method oriented. We describe methods that can be employed for a wide variety of applications, focusing on the principles underlying these methods. Additionally, we discuss similarities and differences among these methods that are independent of any specific application.

1.1   What Is Biomedical Text Mining?

The terms *text mining, literature mining,* or *text data mining* have seen much use within the biomedical domain during the past decade [4, 5, 50, 91, 128, 214, 284]. The general research area denoted by these terms is concerned with making effective use of biomedical text through the application of computational tools.

To properly define *text mining,* we first reexamine the original metaphor giving rise to it, namely, that of mining. We can consider, for example, gold mining, coal mining, or diamond mining and realize at the onset that *text mining* is a misnomer. In all three real-life mining examples, the term *mining* is explicitly modified by the metal or mineral for which we are searching. In contrast, when we use the term *text mining*, we obviously do not imply that we are mining *for* text but rather we are mining *within* the text for something valuable. Because text mining encompasses a broad array of tasks, it is necessary to identify explicit goals along with the text-based methods that may realize them in order to find effective solutions to actual problems.

There are numerous ways in which text can be used to obtain valuable information, and multiple factors determine what may constitute an effective text-based method or a realistic goal for employing it. In the same way that gold mining and coal mining are performed within different geographical regions, using different sets of tools, and targeting different groups of customers, biomedical text-mining applications vary in such aspects as the text sources used, the methods employed, and the users served. For example, the task of searching for textual information about adverse drug reactions due to a specific genetic variation is likely to require different methods than those needed for the task of identifying scientific abstracts discussing gene expression in the mouse.

In broad terms, biomedical text mining aims to identify and present relevant information within scientific text, satisfying the diverse needs of a large community of biomedical scientists. Notably, there is no "average" biomedical client; different users have distinct needs. For instance, clinicians or researchers in a given area may want to find answers in the literature to specific questions such as, "What is the role of DRD4 in alcoholism?" or "How do HMG and HMGB1 interact in hepatitis?"[1] In contrast, bioinformaticians and systems biologists may be interested

---

1. Such information needs were investigated in the 2005 TREC Genomics Track [89].

in finding sentences or paragraphs specifying protein-protein interactions as a first step toward hypothesizing pathways or other molecular networks. Yet another community of potential text-mining clients are biomedical database curators, who may want to rely on automated means to assist them in finding published articles likely to be relevant to the database subject area. For example, FlyBase [245] curators may want a system that retrieves all articles discussing gene expression in the fruit fly *Drosophila melanogaster*, while curators working for the Mouse Genome Informatics resource [158] may require similar papers discussing the expression of genes in the mouse. Other users may be interested in identifying methods, drugs, mutations, or specific diseases. The possibilities are almost limitless.

Given the breadth of users and their respective needs, we need to characterize and define an explicit task when setting out to mine for information within text, considering several important and interrelated aspects:

· *The user*   What is the prospective user's role and what is his or her purpose in seeking the information?

· *The user's information needs*   What questions is the user trying to answer or what biological task is the user aiming to address?

· *The text sources*   Which sources of text are accessible to the user (e.g., public datasets, proprietary patent records, patient records, etc.) and which of these available text sources may be most useful in addressing the user's information need?

· *The representation of the text*   How is the text represented in terms of the native representation (e.g., PDF, XML, or HTML), over which the user typically has very little control, and in terms of the representation that results from processing the native text for an application program that aims to find the needed information?

· *The tools*   Which candidate methods or applications can be employed to satisfy the user's information needs based on the available text and its representation?

· *The evaluation*   Which performance metrics and evaluation procedures are used to measure whether the task was indeed performed successfully?

These questions demonstrate that text mining is indeed not a single task to be realized by a single system; rather, it is a broad domain in which different tasks are addressed using a variety of tools. These tools include, among others, methods from the field of natural language processing—and within it—from the more specific area of *information extraction*. The latter is concerned with the identification of relevant entities and relations within text and the representation of the extracted information in some standard form. Another important task of text mining is to identify the text that is interesting or relevant to the users; this area of text analysis is known as *information retrieval*. Some of the different aspects of the tasks involved in text mining are further illustrated by the example in the next section.

1.2   Example: The *BRCA1* Pathway

To demonstrate some key points in biomedical text mining, we consider the case of the *BRCA1* pathway. It comprises numerous genes, notably including the *BRCA1* and *BRCA2* genes, and is known to be involved in certain types of cancer, especially breast cancer.

In the simplest scenario, a clinician who is starting to study the pathway may be interested in finding information about a certain gene, for instance, the gene *FANCF,* coding for the *Franconi anemia group F* protein. To find information about the gene, the researcher may simply pose a query to PubMed using the term *FANCF* and retrieve a relatively small set of 85 abstracts.[2] The search can be further narrowed to those abstracts containing the words *BRCA1* or *BRCA,* obtaining a smaller set of 18–28 abstracts. The clinician can go through this set of abstracts with relative ease, obtaining and manually summarizing information from it. The measure of success in this case might be as simple as the clinician's satisfaction with the findings. More quantifiable measures may include the ratio of abstracts that the clinician found helpful with respect to the set of retrieved abstracts, or if multiple users have issued a similar query against the system, the average level of satisfaction along a fixed scale (say, a number between 1 and 5) calculated across all users.

In this case, we have demonstrated an information retrieval engine, namely PubMed, as a tool used to effectively obtain the information sought by the user, which was literature about the *FANCF* gene in the desired context. It was straightforward to retrieve relevant documents for the *FANCF* case because this gene name is relatively specific (there are no other human gene names that use this same symbol) and the gene

2. Query issued to PubMed, July 30, 2011.

has only one alternative gene symbol, *FAF*, which is not widely used. However, matters are often not this simple. For many genes, the problems of synonymy (many different names for one gene) and polysemy (the same name referring to multiple genes) complicate the problem of identifying documents that are relevant to a given information need. Additionally, a user's information needs are typically more complex and cannot be formulated as a simple query. Moreover, the results of any potential query to a search engine can include such an immense number of documents that there are too many to be read or analyzed without an additional automated aid.

As an extreme case, suppose that the user is a researcher in the area of systems biology who is interested in *all* the information about the *BRCA1* pathway and has the goal of understanding every known relationship within the pathway and reconstructing it based on everything that is discussed about it in the literature. That is, the user is trying to find out which genes and proteins may be interacting with each other as part of this pathway, what the effects of these interactions are, and what the ultimate role of each gene is at the onset of certain medical conditions. Such a use case was one of the earliest motivating forces for text mining within bioinformatics [20, 67, 109, 129].

As a first step, the user needs to identify the relevant published literature in which all the information about the *BRCA1* pathway can be found. This step itself may already prove to be challenging. For instance, a query using the term *BRCA1* issued against the PubMed database results in 8,241 retrieved abstracts,[3] without even considering any other genes in the pathway or any other relevant sources of information (other databases, lab reports, etc.). To fully understand the pathway, all of these other genes and relevant data sources also need to be queried, resulting in tens of thousands of articles. Notably, in most realistic scenarios, the researcher will not even know at the onset all of the genes involved in the pathway in order to issue queries about them. But even when a gene name is known, as in the example of *BRCA1*, the set of thousands of retrieved documents may still be incomplete. Many potentially relevant articles may not have even been retrieved simply because they were discussing the *BRCA1* gene using one of its synonyms, such as *RNF53*. Moreover, some of the documents retrieved may be irrelevant to the pathway, mentioning the gene in passing without containing a focused discussion of it or of its interactions with other genes.

---

3. Query issued to PubMed on July 30, 2011.

Obviously, even if the first step described above is surmounted, trying to sort through tens of thousands of documents to obtain a coherent picture of the pathway remains a major and not quite well-defined challenge. To automatically obtain information from the text about each gene and protein involved in the pathway, we would need tools to find all mentions of genes and proteins within the retrieved articles, a task referred to as *named-entity recognition*. Automatically identifying potential interactions among the recognized genes and proteins gives rise to an additional and more complicated task known as *relation extraction*. The entity-recognition and the relation-extraction tasks are further complicated by the common use of multiple synonyms to refer to genes and proteins and the multiple ways in which sentences and statements expressed in text may be interpreted, among other ambiguities that are abundant in natural language.

In addition to these issues, another consideration is the need, in some cases, to automatically distinguish between statements that are hedged or speculated and statements that are made with certainty. Moreover, we may want to determine if a given assertion is supported by experimental evidence or not.

1.3   Challenges in Biomedical Text Mining

The example in the previous section described several computational tasks that a biomedical scientist might undertake in order to find out more about *BRCA1*. In subsequent chapters, we discuss these and other tasks in more detail.

Chapter 2 covers several background topics that are key to understanding the discussion of the principal tasks in biomedical text mining that come in later chapters. The first major topic covered is biomedical text sources. That is, what the various resources are that contain the text to be mined. As discussed previously, determining the appropriate source of text is one of the key steps in any biomedical text-mining application. The chapter also covers core methods in natural language processing that support various types of text analysis. The procedures covered here range from segmenting a document into smaller units to predicting the grammatical parse tree for a given sentence. This material is germane to the previously discussed step of determining an appropriate representation for a text-mining task. Additionally, chapter 2 also describes the central challenges that arise in analyzing natural language and discusses the important role of ontologies in biomedical text mining.

One of the challenges confronted in the previous section was the identification of the documents and passages that are relevant to satisfying the information need concerning the *BRCA1* pathway. Chapter 3 covers the tasks of information retrieval (IR) and text categorization. Given a query of interest, such as the name of a gene, the IR task is to identify and return a set of documents that are relevant to the query. Text categorization, which often uses representations and methods similar to IR, is the task of automatically assigning documents to one or more predefined categories of interest.

Another significant challenge that was brought up in the *BRCA1* example was the identification of mentions of the relevant genes and proteins and the extraction of relationships among these entities. Chapter 4 focuses on the task of information extraction (IE). The information-extraction task entails recognizing instances of named entities (e.g., gene and protein names), relations among entities (e.g., protein-protein interactions), and complex events that describe more structured relationships among entities. We describe methods for all three of these tasks in chapter 4.

Chapter 5 discusses how we can empirically evaluate biomedical text-mining systems. In particular, we describe the relevant measures for assessing the performance of text-mining approaches and describe several community-wide evaluations that have been conducted in order to compare competing methods, determine the state of the art, and spur further improvements to the methods.

Chapter 6 describes several systems that bring together methods presented in previous chapters to form valuable applications. Specifically, we cover systems that (1) add value to biological studies by recognizing entities in text sources and linking them to other entities and data resources, (2) support the curation of structured databases, and (3) enable interesting biological predictions and discoveries.

2   Fundamental Concepts in Biomedical Text Analysis

The development of the Internet has made it easy for biologists to create databases and online portals representing various aspects of biological knowledge and to make these resources publicly available. Although there are hundreds of such online resources[1] representing biological knowledge in a structured format, much of the scientific community's knowledge is represented only as unstructured text.

A structured format is one in which information is organized and represented in a formal and predefined manner. For example, a relational database consists of multiple tables corresponding to predefined relations. Each table is defined by a fixed set of fields, each of which has a prespecified meaning and data type. By contrast, information represented in ordinary natural language does not have this structured format. Sentences may describe ideas abstractly or concretely, directly or obliquely. Moreover, sentences describing the same or similar ideas may have very different syntax and employ very different vocabularies.

There are several reasons why a large amount of scientific knowledge is represented only in free-text form. First, most structured databases suffer from a "curation bottleneck." Typically, the contents of these databases are populated and maintained by scientists, known as *curators*, who have expertise in the area covered by the database. The curation bottleneck means that the completeness of a database is limited by the rate at which these curators can find relevant articles, extract the information of interest from them, and enter this information in a structured format into the database. In short, it is often the case that curators cannot keep up with the pertinent literature. Second, there are important facets of the biomedical literature that are not represented by existing databases. For

1. The interested reader is referred to the *Nucleic Acids Research* annual Database Issue, which catalogs many of these systems [72].

example, although there are many databases that characterize the functions of individual genes in various organisms, these databases generally do not describe how the gene functions are disrupted by pathogens that may infect the organisms. Moreover, the nuances and qualifications that typically accompany descriptions of scientific findings in articles are often not represented in structured databases, even when the findings themselves are referenced. For these reasons, there is a compelling need for methods that can exploit the vast web of knowledge that is represented in text sources.

In this chapter, we describe some of the text sources that contain large amounts of biomedical knowledge and discuss the fundamental tasks of *natural language processing* (NLP). Natural language processing is an area that brings together methods from linguistics and computer science in order to automatically analyze and elicit meaning from text that is written in a natural language, such as English. Finally, we discuss the concepts of *controlled vocabularies* and *ontologies* and explain how these concepts are connected to the topic of biomedical text mining.

2.1   Biomedical Text Sources

Biomedical knowledge represented using natural language is found in many online resources. The most accessible source of biomedically relevant text is PubMed [186], which is an online database of journal citations and abstracts. PubMed is managed by the National Library of Medicine, which is part of the US National Institutes of Health (NIH). The largest component of PubMed is a database called MEDLINE that indexes more than 5,000 biomedical journals on a regular basis. In addition to citation information and abstracts, MEDLINE entries also include index terms from a controlled vocabulary called Medical Subject Headings (MeSH), which is discussed in section 2.5. PubMed is a superset of MEDLINE that includes other citations, such as articles that are awaiting MeSH indexing before being included in MEDLINE, articles that were published in a given journal before it was selected for inclusion in MEDLINE, and other special cases.

The PubMed portal [186] provides links to the full text of many articles and a variety of other services such as references to similar articles and the ability to save and automatically update queries, among others.

Another obvious source of biomedically relevant text is the primary scientific literature. Virtually every biological journal has an associated website that makes published articles available in electronic form. For many of these journals, access to the full text of the articles requires either having an institutional subscription to the journal or paying a per-article fee. Commonly, online journal articles are published in PDF and HTML formats, and thus the text of the articles is available for automatic analysis, after some processing to extract it from the respective PDF or HTML files.

In addition to publishers' websites, many articles are also available at the PubMed Central portal [187]. PubMed Central is a distinct entity from PubMed but is also managed by the National Library of Medicine. PubMed Central provides free access to many articles published in the biomedical and the life sciences literature. Articles that are the result of NIH-funded research are now required to be deposited into PubMed Central.

A large number of biomedical text-mining studies have focused solely on processing abstracts from PubMed. There are several advantages to working with abstracts: they are freely available and easily downloaded, and they tend to be rich in information content because they summarize the main points of their associated articles. Moreover, being concise, they are less challenging to computationally analyze than full-text articles. However, because only part of the information in any article is recapitulated in the abstract, there is increasing interest in processing the full text of available articles, including figures [42, 168, 212] and captions [1, 196].

In addition to repositories of scientific articles and abstracts, such as PubMed, text is often found in selected fields of databases that largely contain structured information. Consider, for example, the UniProtKB/ Swiss-Prot database [248], which is a leading source of information about proteins. Each protein in the database is described by fields that provide, among other information, the amino-acid sequence, functional features in the sequence, taxonomic information, and names of the protein. Many of these fields are highly structured in that their contents can be unambiguously parsed and processed by simple, automated routines. The CELLULAR COMPONENT field, for example, contains terms from a predefined vocabulary denoting compartments within a cell, and the PROTEIN NAMES field contains strings that define names for the proteins. Some of the fields, however, are filled with unstructured text descriptions. For example, the FUNCTION field for the yeast protein *UBC6* contains the following text:

Catalyzes the covalent attachment of ubiquitin to other proteins. Seems to function in the selective degradation of misfolded membrane proteins in the endoplasmic reticulum.

The content of such fields may be useful for many types of analyses. However, text-mining methods are needed in order to extract the desired information from them.

2.2   Natural Language Concepts

Clearly, there is an abundance of biomedically relevant text available in a variety of online repositories. The units of text contained in these data sources span a range from sentence fragments to multipage articles. Despite the variation in the size and the nature of these text "documents," they share the common property that computational methods for natural language processing are required in order to effectively use and elicit meaning from them. In this section, we describe some of the key aspects of natural language that are taken into account by such computational methods.

Consider the two sentences shown in figure 2.1 taken from the abstract of an article about the previously mentioned yeast protein *UBC6*. Some understanding of the sentences can be gained by considering their orthography. *Orthography* is defined as the conventions of a writing system that are used to express sentences in a given language. Orthography includes aspects of writing such as spelling, punctuation, word breaks, and capitalization. Each sentence in figure 2.1 can be viewed as a sequence of *tokens*, where each token is a word, a word-like sequence of characters, or a punctuation mark. Orthographic information tells us something about how words are grouped into sentences or phrases. Specifically, punctuation tokens indicate phrase and sentence boundaries, and capitalized words often indicate the beginning of sentences. Orthographic information may also provide evidence about the types of entities represented by a given token. The token *UBC6,* for example, contains numeric characters and is capitalized, suggesting that it might be the name of a protein. As is the case with most types of linguistic information, these orthographic cues are often ambiguous. For example, a period typically indicates the end of a sentence but sometimes denotes an abbreviation, as in the species name *S. cerevisiae* (which is shorthand for *Saccharomyces cerevisiae*). Similarly, capitalization sometimes indicates a proper noun, as with *UBC6*, instead of the start of a sentence.

The *morphology* of words is another linguistic aspect that carries information about their meaning. Morphology is the description of the internal structure of words and their formation from smaller units, called

Here we report the identification of an integral membrane ubiquitin-conjugating enzyme.
This enzyme, UBC6, localizes to the endoplasmic reticulum, with the catalytic domain facing the cytosol.

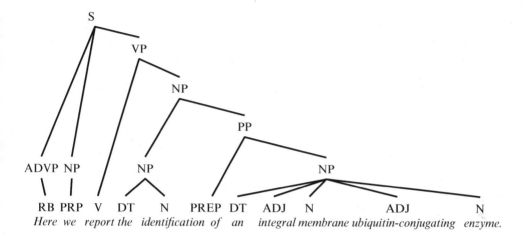

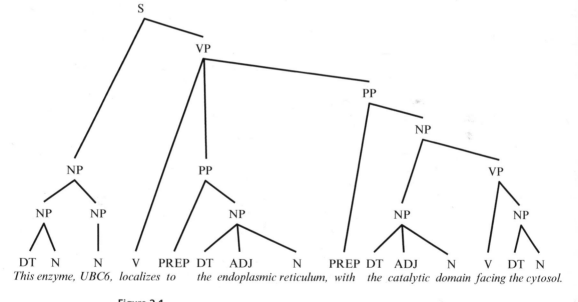

Figure 2.1
Two sentences from an abstract (PMID 8396728) along with their respective parse trees.
The parse trees show the part-of-speech tags for each word and the grouping of words into
phrases. The phrase tags ADVP, NP, PP, and VP represent adverb phrases, noun phrases,
prepositional phrases, and verb phrases, respectively.

morphemes, that are often common to several words. As an example, consider the word *localizes* in the first sentence in figure 2.1. One unit in this word is *local-,* which is itself a word and a common component of other words such as *locality, localization,* and *localize.* Similarly, the word *enzyme* contains the unit *enzym-,* which is a common component of *enzymatic, enzymology,* and *enzymes.* Many morphemes function as prefixes or suffixes that are attached to other word units. In the previous example, the suffix *-ology,* meaning the "study of," was attached to *enzym-* to make *enzymology.* Commonly used prefixes include *pre-, re-,* and *un-*; commonly used suffixes include *-ing, -ly,* and *-omics.* As all of these examples illustrate, morphological analysis can reveal meaningful similarities among words, and in some cases can provide clues about the type of entity that is denoted by an unrecognized word. For example, enzyme names typically end with the unit *-ase,* and thus we may be able to determine that a given name refers to an enzyme even if we have not previously seen the name.

The aspects of language that we have discussed up to this point are focused on each word in isolation. To fully understand the meaning of a sentence, we typically need to elucidate the relationships among its words. One step in this direction is to determine the *part of speech* (POS) of each word. Part of speech denotes the syntactic role of a word in a given sentence. These roles can be grouped into a small number of categories. In figure 2.1, the part of speech is indicated above each word. In the second sentence, *This* is a determiner, *enzyme* and *UBC* are nouns, *localizes* is a verb, and so on. The set of part-of-speech categories shown in figure 2.1 is fairly simple but there are finer-grained POS systems. The Penn Treebank Project [150], which has assembled large corpora of manually curated sentence-structure annotations, employs a system of 36 different POS tags.

The words that compose a given sentence are not ordered arbitrarily but rather are grouped into phrases that have a regular internal structure and relationships linking them together. For example, noun phrases contain one or more nouns, and optionally, determiners and adjectives. One of the noun phrases in the second sentence in figure 2.1 is "*the endoplasmic reticulum,*" which is composed of the determiner *the,* the adjective *endoplasmic,* and the noun *reticulum.* Verb phrases contain a verb along with the constituents of the sentence that depend syntactically on this verb, such as direct objects. A *prepositional phrase* consists of a preposition followed by a noun phrase. The tree structures in figure 2.1 show how the words in the sentences are grouped into phrases and how phrases themselves are constituents in more extensive phrases. The root of each tree represents the entirety of the sentence. As we discuss in the following section, there are various schemes for representing dependencies among words and phrases in a sentence.

Typically, we cannot grasp the full meaning of a passage by considering each sentence in isolation. Instead, relationships among the sentences are key to understanding the passage. Consider the second sentence in figure 2.1. This sentence starts with the words, "*This enzyme, UBC6*," which is a reference to the enzyme described in the previous sentence. From this reference, we can infer that *UBC6* is an integral, membrane ubiquitin-conjugating enzyme, even though the name of the protein and its role are described in separate sentences. More generally, there are many relationships that link together sentences in any given passage. The analysis of relationships among sentences is referred to as *discourse processing* in linguistics.

2.3   Challenges in Natural Language Processing

Now that we have discussed some of the linguistic concepts that are central to natural language processing, let us consider the challenges faced in trying to automatically analyze and elicit meaning from documents. Here we assume that we have already identified and gathered the documents to be processed; in chapter 3, we discuss the task of identifying documents that are relevant to a given information need. In a nutshell, the key problem that arises in understanding natural language (for both automated methods and humans) is *ambiguity*. Typically, linguistic units and relationships have multiple possible interpretations. Here, we consider a few specific manifestations of this general phenomenon of ambiguity.

The first aspect of natural language we discussed in the previous section is orthography, and we pointed out that there may be ambiguity in what certain symbols and typographic conventions indicate. Periods sometimes denote the ends of sentences but sometimes indicate abbreviations. Hyphens are sometimes used to join distinct words but are sometimes used as part of a single word, as in some gene names (e.g., *Actin-5C*).

The meaning of an individual word is sometimes ambiguous as well. Here we consider three classes of such words. *Homonyms* are words that share the same spelling and pronunciation but have different, unrelated meanings. For example, *left* can be an adjective indicating the opposite of *right* or a verb denoting the past tense of *leave*. *Heteronyms* are words that have the same spelling but different pronunciation and meaning. The word *minute,* for example, can be a noun referring to a unit of time or an adjective describing something as very small. The pronunciation of *minute* varies depending on the intended meaning. In contrast to homonyms and heteronyms, *polysemes* are words that are spelled identically but have related, albeit slightly different, meanings. For example, the word *assembly* can mean either the process of something coming together or the end result of this process. One common case of polysemy that occurs in the biomedical literature is that many gene symbols/names refer to more than one gene. Moreover, in many cases, the same name may refer to a gene, the RNA, or the protein product of the gene.

A related phenomenon is *synonymy*, which refers to different words that have identical or very similar meanings. In the biomedical literature, synonymy, like polysemy, is a common issue with gene symbols and names. That is, many genes are referred to by multiple names. For example, the yeast gene *UBC6* is also known as *DOA2.* When multiple passages of text from various sources use different names to refer to the same gene, it can be challenging to determine which passages are referring to the same object.

The challenges we have discussed up to this point are focused primarily on analyzing each word in isolation. The foremost difficulties in natural language processing, however, stem from syntactic ambiguity. By syntactic ambiguity we mean that the relationships among the words in a sentence are not straightforward to discern. For example, there are two syntactically valid interpretations of the phrase "*the MS and blood pressure methods*" [81]. In one interpretation, the word *MS* is an adjective modifying *methods* (i.e., *MS* and *blood pressure* both refer to methods) but in the alternative interpretation *MS* is a noun (i.e., *blood pressure methods* refers to a set of methods, but *MS* is not a method).

Another type of syntactic ambiguity pertains to the attachment of prepositional phrases. Consider the prepositional phrase "*with the catalytic domain facing the cytosol*" in the second sentence in figure 2.1. The catalytic domain mentioned in this case could be part of the subject of the sentence ("*the enzyme, UB6*") or it could be part of the direct object of the sentence ("*the endoplasmic reticulum*"). Although these two interpretations would result in different parse-tree representations of the sentence, either one would be syntactically valid. In general, human readers resolve syntactic ambiguity by exploiting knowledge of the topic being discussed. Knowing that proteins have subunits called domains and that enzymes perform catalytic functions makes it clear to a human reader that the catalytic domain is a part of *UBC6* in this particular example.

Earlier we mentioned how the phrase "*This enzyme*" in the second sentence of figure 2.1 refers to the same entity as *UBC6* in the following phrase and "*integral membrane ubiquitin-conjugating enzyme*" in the preceding sentence. Such instances in which a word or a phrase refers to something described in another phrase or sentence are called *co-references*. As this case indicates, co-references can be manifested in various ways. The phrase "*This enzyme*" is an example of a definite noun phrase that refers to a previously mentioned noun. The phrase "*integral membrane ubiquitin-conjugating enzyme*" is an example of an appositive, in which a noun phrase defines or renames another noun phrase that is adjacent to it. Co-references can also be indicated by pronouns such as *he* or *it* and by proper noun phrases, as in the following case in which *Parkinson's* at the end of the sentence is co-referent with *Parkinson's disease* near the beginning:

A diagnosis of Parkinson's disease is based on the presence of symptoms, some of which occur in elderly people who do not have Parkinson's.

As one might expect, co-references represent another potential source of ambiguity. Consider, for example, the following pair of sentences [175]:

By comparison, PMA is a very inefficient inducer of the jun gene family in Jurkat cells. Similar to its effect on the induction of AP1 by okadaic acid . . .

Here the pronoun *its* could potentially refer to *PMA*, *the jun family*, or *Jurkat cells*. Human readers (or at least readers who are experts in the subject domain) are usually able to resolve such ambiguities by using their knowledge of the world and the specific domain being discussed. In this example, because *PMA* is described as having the role of an inducer in the first sentence, and the entity referred to as *its* in the second sentence is described in terms of its inducer role, we can infer that *its* refers to *PMA*.

A final source of ambiguity we consider is quantifier scope ambiguity. A quantifier is a word that specifies the number or the amount for some entity of interest. Some examples of quantifiers in English are *all*, *some*, *many*, and *three*. The sentence "*Three ligands bind to every member of the family*" contains two quantifiers, *three* and *every*. This sentence could have two possible interpretations: (1) three particular ligands bind to all members of the family or (2) each member of the family binds three ligands, but possibly different ligands for each member of the family.

2.4 Natural Language Processing Tasks

As we have seen, meaning is conveyed at various levels in natural language—through orthography, morphology, syntax, and discourse. Moreover, there are numerous sources of ambiguity that make it a challenging task to automatically elicit the meaning of a passage of text. Nevertheless, computational linguists have developed a powerful toolbox of algorithms for analyzing natural language [115, 148]. In this section, we describe several key tasks that are performed by such algorithms.

Often, the first task that must be addressed before any other processing can be performed is to segment the documents of interest into smaller units. At the coarsest level of granularity, we might divide a document into distinct sections, figures, tables, captions, and footnotes. Within these units, we can segment passages of text into paragraphs and sentences. And within sentences, we can segment the text into distinct tokens. Although these segmentation steps may seem straightforward, each poses challenges.

Consider the task of segmenting a document, such as a scientific article, according to its organization into title, author, abstract, section, and subsection components. The difficulty of this segmentation depends on the format in which the document is represented. Articles in PDF format can be difficult to segment because PDF is intended to represent the presentation structure of the document (i.e., how it will look) rather than its organizational structure. Several methods have been developed for automatically reconstructing the organizational structure of PDF articles [98, 249], although these methods generally require that some information about journal-specific formatting conventions be provided. By contrast, the organizational structure of an article is often easier to recover from an HTML document because the HTML tags themselves may implicitly or explicitly indicate the structure of the document. As in the case of PDF, however, some journal-specific knowledge is still usually required to interpret these tags correctly. In addition to segmenting a given document, processing of the PDF or HTML source may also detect and retain information about stylistic features of particular passages, such as which tokens are displayed in italics, bold, subscripts, and superscripts.

There are also applications in which the figures and captions in articles are of particular interest. In such applications, figures and captions must be delineated from the other parts of the article and each caption must be paired with its corresponding figure [167].

In addition to segmenting documents at a coarse level of granularity, for some tasks it is necessary to perform finer-grained segmentation, such as recognizing sentence boundaries. Sentences usually end with a period, question mark, or exclamation point. Thus we could use occurrences of these characters to break text into sentences. However, as noted, a period does not necessarily indicate the end of a sentence. In addition to a sentence boundary, a period can signal a decimal point, an abbreviation, or even an abbreviation at a sentence boundary.

Similarly, consider the difficulties in recognizing token boundaries in biomedical text. This task is referred to as *tokenization*. Given the sentence boundaries, one naïve approach is to split sentences into tokens by using white-space characters as token boundaries. In many cases, however, we may want to treat other characters as token boundaries. For example, it makes sense to use the hyphen as a token boundary in the phrase "*analysis of peptide-protein interactions*," and parentheses as token boundaries in the phrase "*heat response genes, including OLE1(YGL055W)*." However, the gene/protein names *(MIP)-1alpha*, *pRB/p105*, and *TrpEb_1* demonstrate cases in which we do not want to use punctuation characters as token boundaries because doing so would result in individual entity names being erroneously broken into multiple tokens.

Methods for sentence segmentation [273] and tokenization [111] generally involve using sets of heuristic rules to recognize the relevant boundaries. To segment sentences, for example, we might use a rule specifying that multiple periods inside of brackets or parentheses indicate abbreviations, not the ends of sentences. Another rule might be that when a word is followed by a period and then a comma, the period denotes an abbreviation.

The objective of the tokenization process can vary depending on the application in which it is used. For an application such as information extraction (discussed in chapter 4), in which the sentences are processed by a parser, we would like the tokenization process to preserve important punctuation in the sentences. Elements of punctuation, such as commas and parentheses, provide key information to the parser, and thus we want them to be preserved and explicitly represented as part of the tokenization output. For other applications, such as ad hoc information retrieval (discussed in chapter 3), it might be acceptable for the tokenizer to remove some characters. The task performed by ad hoc retrieval systems is to identify documents that match or are similar to a given query. Usually punctuation is not relevant to determining such matches or similarities.

After tokenization, further processing may be done to reduce the resulting tokens into *terms*. To explain the concept of a term we first describe the distinction between *tokens* and *types* [147]. Whereas a type is a specific sequence of characters, a token is a specific occurrence of a corresponding type. In the sentences in figure 2.1, for example, there are two tokens corresponding to the type *enzyme* (the last word in the first sentence and the second word in the second sentence) and four tokens corresponding the type *the*. For many text-mining applications, such as the information-retrieval systems discussed in chapter 3, documents are represented by a set of terms. Some of these terms directly correspond to the types referenced in the document but others may be derived by normalizing types or forming combinations of them.

Perhaps the simplest form of normalization is *case normalization,* in which all alphabetical characters are mapped to their lowercase equivalents. For example, case normalization would map *Enzyme* and *enzyme* to the same term. Typically, we think of the lowercase variant (*enzyme* in this case) as being the standard one. In some applications, case differences are not important and thus case normalization is a sensible thing to do. However, in other contexts, case information may convey important semantics. For example, consider the three words *tagA*, *TagA,* and *taga*. The first of these tokens is the name of a gene, the second is the name of the corresponding protein, and the third is a four-character DNA sequence. For some species, it is common to distinguish a protein name from its counterpart gene name by capitalizing the first letter in the protein's name. Given that all its letters appear in the same case, *taga* probably does not refer to a gene or protein name but instead to a sequence of nucleotides.

Another class of normalization processes reduces related words to a common root form. For example, the words *modify, modifies,* and *modifying* are different forms of the same verb. For some applications, we may want to collapse these distinct words into the same term in order to make their relatedness explicit. For example, when considering the similarity of two documents, we might want the presence of the word *modify* in one document and *modifying* in the other to be counted as evidence for their similarity. There are two types of methods for mapping related words to a common term, *stemming* and *lemmatization*.

Stemming is a process that uses a set of heuristic rules to selectively trim the ends of words. The most widely used method for automatically stemming English words is Porter's algorithm [182]. Given the words *modify, modifies,* and *modifying,* Porter's algorithm reduces all three to the root form *modifi-*. In this example, *modifies* is stemmed by a rule that specifies that an *ies* ending should be reduced to *i*. The Porter stemmer sequentially applies five phases of word reductions, with each phase employing a set of transformation rules along with conditions on when each rule can be applied. Although stemming often results in sensible reductions of words, it sometimes fails to reduce related words to the same root and other times mistakenly does so. For example, the Porter stemmer reduces both *experiment* and *experience* to *experi*. Moreover, its rules do not recognize the relationship between *mouse* and *mice*. The former is reduced to *mous* and the latter is left unchanged.

Lemmatization employs an idea similar to stemming but is more linguistically oriented. Whereas a stemmer operates on an individual word without taking into account its context, a lemmatizer may consider the part of speech of a word, morphological rules, and a dictionary of base forms, known as *lemmas*. In contrast to a stemmer, a lemmatizer would recognize that both *mouse* and *mice* are forms of the lemma *mouse*.

The particular normalization processes that are used in text-mining systems vary depending on the application. Some applications may not involve any normalization, whereas others employ all of the normalization procedures we have discussed.

In addition to terms that are derived through the various normalization processes we have discussed, we may also have terms that represent combinations of words. The most common such terms are *n-grams*, which represent specific occurrences of *n* consecutive words. For example, the phrase "*with the catalytic domain facing the cytosol*" includes the following bigrams (*n* = 2):

with the
the catalytic
catalytic domain
domain facing
facing the
the cytosol

and the following trigrams ($n = 3$):

with the catalytic

the catalytic domain

catalytic domain facing

domain facing the

facing the cytosol.

Including *n*-grams in a representation enables some information about local word dependencies to be captured. For example, if we are interested in documents pertaining to the nuclear membrane, we can search for such documents using the bigram *nuclear membrane*. This bigram would identify a more specific and relevant set of documents than simply searching for documents that contain both the words *nuclear* and *membrane* because this latter search does not specify the order and immediate adjacency of the two words. When not occurring consecutively, *nuclear* and *membrane* can refer to other nuclear structures and other types of membranes.

We have described here only some of the most common types of terms. Depending on the application domain and the genre of text, many other term variants can be used, incorporating ideas such as normalization of numbers, correction of common misspellings, and the extraction of character *n*-grams from words [155].

Another type of processing that is useful for many applications is part-of-speech (POS) tagging. As discussed earlier in the chapter, the part of speech associated with a word refers to its role in a given sentence. Part-of-speech tagging is the process of automatically predicting a POS tag for each token in a given document. The tags assigned by such a procedure are selected from a predefined set. One commonly used set is the Penn Treebank Project POS tags. This set of 36 tags describes parts of speech at a finer level of granularity than is shown in figure 2.1. For example, the Penn Treebank uses the following four tags for categorizing nouns:

NN   noun, singular or mass

NNS   noun, plural

NNP   proper noun, singular

NNPS proper noun, plural

Most modern part-of-speech taggers are based on models that are induced by supervised machine-learning methods from labeled corpora [29, 193, 252]. A labeled corpus, in this case, consists of sentences in which linguists have manually indicated the correct POS tag for each token. Given such a corpus, machine-learning methods are able to identify statistical regularities in the context of a given token that enable its part of speech to be disambiguated. Although there are general-purpose POS taggers available, it has been shown that taggers that are specifically trained on biomedical corpora are more accurate when applied to biomedical documents [224, 254].

The next step of syntactic analysis, beyond POS tagging, is to perform *chunking* [116] or *parsing* [81]. Chunking is the process of grouping tokens into grammatical phrases. Parsing is the process of analyzing a sequence of tokens in order to determine its structure with respect to a formal grammar.

A simple form of chunking is to group words into base noun phrases. A base noun phrase is a noun phrase that is nonrecursive; that is, it does not contain another noun phrase. For example, when applied to the first sentence in figure 2.1, an accurate chunker for identifying base noun phrases would recognize "*the identification*" as one base noun phrase and "*an integral membrane ubiquitin-conjugating enzyme*" as another. More ambitious forms of chunking involve recognizing other types of phrases, such as verb phrases or even fully partitioning sentences into chunks of various predefined types.

In contrast to chunking, parsing means mapping sentences into a representation that more fully describes the grammatical relationships among the constituents of the sentence. A *parser* is a software system that performs this mapping. Broadly speaking, there are two ways in which parsers represent the organization of sentences. One type is a *phrase structure* organization, as shown in figure 2.1. In this representation, which is called a *parse tree,* words are organized into nested phrases, with the top level encompassing the whole sentence.

The other way in which sentence organization can be represented is as a *dependency structure*. Figure 2.2 shows the dependency structure of the first sentence from figure 2.1. The dependency structure indicates binary, grammatical relations between pairs of words. Consider, for example, the top part of figure 2.2. There are three grammatical relations between the verb in the sentence, *report,* and other words: *Here* is an adverbial modifier (ADVMOD) of *report, we* is the nominal subject (NSUBJ) of the verb, and *identification* is the direct object (DOBJ) of the verb.

Although parsing provides more information about a sentence than chunking, there are cases in which chunking is preferred, because full parsing is computationally more expensive than chunking and it is not

Here we report the identification of an integral membrane ubiquitin-conjugating enzyme.

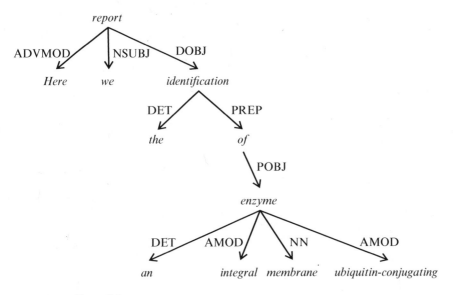

Figure 2.2
A dependency parse for the first sentence from figure 2.1. The arrows represent specific grammatical relations between word pairs, and the label next to each arrow indicates the type of relation.

as robust. That is, a parser has to resolve more ambiguity than a chunker and is thus less likely to return the correct representation of a given sentence. For many applications the representation provided by a chunker is sufficient.

Most modern systems that perform parsing and chunking are based on models learned from labeled data. Parsers are trained using *treebanks*,[2] which consist of sentences that have been manually annotated with correct parse structures by linguists. The Penn Treebank [150] is one such resource that has been widely used for parser development and evaluation. This treebank consists mostly of newswire stories and other nonbiomedical sources of text. The accuracy of a parser, however, is sensitive to the problem domain and the genre of text being parsed [153].

---

2. Chunkers are often trained on treebanks as well, although they do not require annotations that are as detailed as parsers do.

For this reason, specialized treebanks, such as GENIA [242], have been developed for the biomedical domain, and specialized parsers and chunkers have been trained on these treebanks [116, 153]. Although automatic parsers and chunkers are not able to produce syntactic analyses of sentences with perfect accuracy (in any problem domain), these syntactic analyses have nevertheless proven useful for many applications.

Another type of analysis that can provide value for some biomedical text-mining applications is the recognition of statements that express negation and speculation. An example of a statement expressing negation is *"TRADD did not interact with TES2,"* and an example of a speculative statement is *"Pdcd4 may thus constitute a useful molecular target for cancer prevention."*

Although negation is a fairly straightforward concept, the notion of speculation is not as simple to articulate. One analysis of speculation in the biomedical literature [156] has identified several different categories of speculative statements. Two examples of the categories of speculative statements defined in this study are statements of speculative hypotheses (e.g., *"To test whether the reported sea urchin sequences represent a true RAG1-like match . . ."*) and statements of knowledge paucity (e.g., *"How endocytosis of Dl leads to the activation of N remains to be elucidated."*).

The tasks of recognizing speculative or negated statements can be addressed at various levels of granularity. The simplest approach is to classify sentences according to whether they contain a speculative (negated) statement or not [156]. An alternative approach for speculation is to classify sentences according to the degree of speculation expressed. For example, Light et al. [143] classified sentences into three categories: *low speculative, high speculative,* and *definite.* The notion of a low-speculative statement is one in which the authors posit something that follows almost directly from the results but not quite. Other investigators have developed an approach in which sentence fragments are simultaneously classified along multiple dimensions, including polarity (positive versus negative statement), level of certainty, and others [216].

A more detailed approach to recognizing speculative or negated statements is to identify a cue for each statement and to determine the scope of the negation or speculation. Consider, for example, the following speculative sentence [64]:

Although IL-1 has been reported to contribute to Th17 differentiation in mouse and man, it remains to be determined <u>whether therapeutic targeting of IL-1 will substantially affect IL-17 in RA</u>.

Here the word *whether* is a cue indicating that what follows is speculative and the underlined part of the sentence corresponds to the scope of the speculation. The relationship between *IL-1* and *Th17* differentiation is not described as being speculative in this sentence, and therefore it is not part of the scope of the speculation. As with POS tagging, chunking, and parsing, the state-of-the-art methods for detecting speculation in biomedical text are based on learning models from labeled training data [64].

2.5   Biomedical Vocabularies and Ontologies

Up to this point, the discussion in this chapter has focused on (1) sources of biomedical text, (2) key issues that arise in representing and analyzing it, and (3) methods for such analysis. In this section, we turn our attention to a different source of biomedical knowledge that has several important relationships to text. In particular, we discuss the interrelated concepts of *controlled vocabularies, taxonomies, thesauri* and *ontologies*. All of these concepts refer to ways in which knowledge about a particular domain can be represented formally.

A controlled vocabulary is a set of terms that have been carefully selected to characterize some domain of interest. The set of terms in the vocabulary provides a standardized way of indexing documents. This set is typically chosen to provide nearly complete coverage of the domain of interest while minimizing redundancy.

A taxonomy, by contrast, organizes a controlled vocabulary into a hierarchical structure. Each term in a taxonomy has one or more parent-child relationships with other terms in the taxonomy. These parent-child relationships may represent the IS-PART-OF relation (e.g., M phase IS-PART-OF cell cycle), IS-A relation (e.g., meiotic cell cycle IS-A cell cycle), or possibly other relations. A thesaurus enhances a controlled vocabulary by representing nonhierarchical relationships among terms. For example, these relationships may encode which terms are synonyms or near synonyms of each other.

The most prominent controlled vocabulary in the biomedical domain is the collection of Medical Subject Headings (MeSH) [145, 157], which has been developed by the National Library of Medicine. Since MeSH encodes hierarchical and other relationships among terms, it can be considered to be a taxonomy and a thesaurus.

The core of MeSH is a set of terms called *descriptors* or *main headings*. The primary role of descriptors is to index citations in the MEDLINE database. The descriptors are arranged hierarchically from most general to most specific. The top level of the hierarchy defines 16 different categories, including anatomic terms, organisms, diseases, drugs, and so on. Each MeSH descriptor appears in at least one place in the hierarchy and may appear in as many additional places as appropriate.

The meaning of the relationship between a parent descriptor and child descriptor in MeSH is not precisely defined. The relationship has been described as representing "broader than" which, in practice, should mean that the set of documents returned in response to a query on the child descriptor should be nearly a subset of the documents returned for a query on the parent descriptor.

Another component of the MeSH vocabulary is a set of qualifiers, also known as *subheadings,* which are used in conjunction with descriptors. There are 83 qualifiers that represent commonly used refinements of descriptors.

Additionally, each descriptor can have one or more entry terms that are synonyms, near synonyms, or otherwise closely related to the descriptor.

Table 2.1 shows some of the information that MeSH encodes for the descriptor EMBRYONIC DEVELOPMENT. The **Entry Terms** field lists other terms that are synonyms, near synonyms, or otherwise closely related to

Table 2.1
Several fields associated with the MeSH descriptor EMBRYONIC DEVELOPMENT

MeSH heading	EMBRYONIC DEVELOPMENT
Tree Numbers	G07.700.320.500.325.180
	G08.686.785.760.170.104
Scope Note	Morphological and physiological development of embryos
Entry Terms	EMBRYO DEVELOPMENT
	EMBRYOGENESIS
	EMBRYONIC PROGRAMMING
	POST-IMPLANTATION EMBRYO DEVELOPMENT
	POSTIMPLANTATION EMBRYO DEVELOPMENT
	POSTNIDATION EMBRYO DEVELOPMENT
	POSTNIDATION EMBRYO DEVELOPMENT, ANIMAL
	PRE-IMPLANTATION EMBRYO DEVELOPMENT
	PREIMPLANTATION EMBRYO DEVELOPMENT
	PRENIDATION EMBRYO DEVELOPMENT, ANIMAL
Allowable Qualifiers	DRUG EFFECTS
	GENETICS
	IMMUNOLOGY
	PHYSIOLOGY
	RADIATION EFFECTS
Unique ID	D047108

Table 2.2
One of the MeSH tree structures encompassing the descriptor EMBRYONIC DEVELOPMENT

PHENOMENA AND PROCESSES [G]
  PHYSIOLOGICAL PHENOMENA [G07]
    PHYSIOLOGICAL PROCESSES [G07.700]
      GROWTH AND DEVELOPMENT [G07.700.320]
        MORPHOGENESIS [G07.700.320.500]
          EMBRYONIC AND FETAL DEVELOPMENT [G07.700.320.500.325]
            ECTOGENESIS [G07.700.320.500.325.089]
            EMBRYONIC DEVELOPMENT [G07.700.320.500.325.180]
              CELL LINEAGE [G07.700.320.500.325.180.500]
              EMBRYONIC INDUCTION [G07.700.320.500.325.180.750]
              GASTRULATION [G07.700.320.500.325.180.812]
              NEURULATION [G07.700.320.500.325.180.875]
            FETAL DEVELOPMENT [G07.700.320.500.325.235]
            ORGANOGENESIS [G07.700.320.500.325.377]
            SEX DIFFERENTIATION [G07.700.320.500.325.520]

Another tree (not shown) containing EMBRYONIC DEVELOPMENT is rooted at the descriptor
REPRODUCTIVE and URINARY PHYSIOLOGICAL PHENOMENA.

the descriptor. The **Allowable Qualifiers** field lists the qualifiers that can
be used to modify the descriptor. For example, an article about genetics
of embryonic development would be indexed by the descriptor/qualifier
pair EMBRYONIC DEVELOPMENT/GENETICS. The **Tree Numbers** field indicates that the descriptor is located in two different places in the hierarchy.
Table 2.2 depicts one of these trees and indicates the location of EMBRY-
ONIC DEVELOPMENT within it.

In addition to MeSH, there are other controlled vocabularies that
are used to index the biomedical literature. The Unified Medical Language System (UMLS) [144, 256] is a set of resources developed by the
National Library of Medicine for linking together and manipulating
concepts that are represented in a compendium of more than 100 source
vocabularies.

A concept closely related to those just discussed is that of an ontology,
which provides explicit formal specifications of the terms of interest and
relations that exist among them in some domain of interest. Although
the distinction between an ontology and a controlled vocabulary, taxonomy, or thesaurus is neither precisely defined nor widely agreed on,
there are several properties that generally characterize ontologies. First,
typical ontologies explicitly describe four types of concepts: *instances,
classes, attributes,* and *relations*. Instances are the ground-level objects
represented in the ontology, and classes describe sets of instances. Relations define the relationships that can exist among instances and among
classes. Attributes characterize properties of instances, classes, or rela-

tions. Additionally, an ontology may also include logical axioms that specify inferences that can be made about objects represented in the ontology. Moreover, ontologies generally are represented in a formal logic-based language and have precise semantics specifying the meaning of the terms represented. There are many biomedical ontologies that do not have all of these characteristics but are nevertheless considered to be ontologies. The most essential characteristics are representation in a formal language and precise semantics.

The most widely used biomedical ontology is the Gene Ontology (GO) [246, 247], which has been used to specify millions of annotations describing the functions of hundreds of thousands of gene products. The Gene Ontology consists of three constituent ontologies that together contain approximately 30,000 terms. The CELLULAR COMPONENT ontology describes the parts of a cell or its extracellular environment. The MOLECULAR FUNCTION ontology describes the activities of a gene product at the molecular level, such as binding or catalysis. The BIOLOGI-CAL PROCESS ontology details processes involved in the functioning of cells, tissues, organs, and organisms. Relationships among the terms are expressed using several different relations: IS-A, PART-OF (and its inverse, HAS-PART), REGULATES, POSITIVELY-REGULATES, and NEGATIVELY-REGULATES. Figure 2.3 illustrates a few GO terms and the relationships among them.

Although the Gene Ontology does not incorporate all of the characteristics of ontologies discussed previously, it does possess the most critical of these, having a formal representation with precise semantics. For example, in contrast to MeSH, which uses the loosely defined BROADER-THAN relation to indicate how a parent term relates to a child term, the Gene Ontology uses the precisely defined set of relations mentioned above.

As mentioned earlier, there are numerous other ontologies that pertain to biomedical domains, and there is an ongoing effort, the Open Biomedical Ontologies (OBO) consortium, focused on facilitating interoperability among a broad set of biomedical ontologies [223]. To ensure interoperability, the OBO consortium specifies a number of requirements that ontologies must meet to be considered as OBO Foundry ontologies. For example, these requirements specify that constituent ontologies use a common syntax and that each term has an identifier that uniquely identifies it across the ontologies.

We have discussed the concepts of controlled vocabularies and ontologies without yet stating why they are relevant to the enterprise

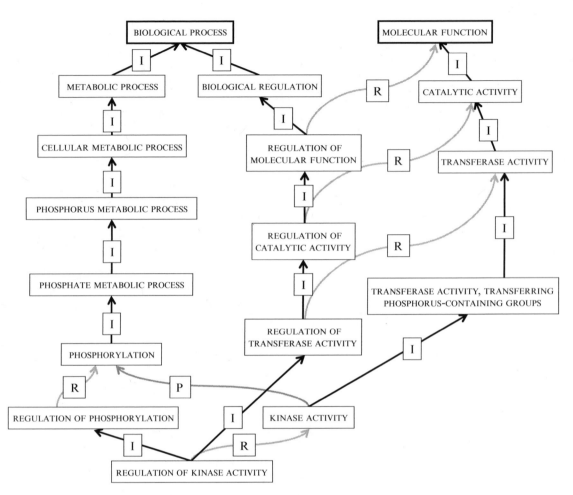

Figure 2.3
Some Gene Ontology terms and the relations among them. Is-A relations are denoted
with an I, REGULATES relations are denoted with an R, and PART-OF relations are
denoted with a P.

of biomedical text mining. We now turn our attention to this question by considering several distinct connections between ontologies[3] and biomedical text mining. The first connection is that a central goal of text mining is to elicit semantic representations from natural language, and ontologies provide a framework for defining semantic concepts and their inter-relationships. Thus, as we discuss in chapter 4, an ontology can be used to specify the set of concepts and relations that are used to describe knowledge that is automatically extracted from text.

A second connection is that text-mining approaches may be useful in helping to refine ontologies. For example, Ogren et al. [176] conducted an analysis in which Gene Ontology terms were compared in order to reveal their compositional structure. They found that certain substrings, which themselves do not correspond to GO terms, recur frequently in the ontology. These recurring substrings often represent interesting semantic relationships among the related terms. One such recurring substring they identified was *regulation of*. There are many pairs of terms in which one term is related to another by the addition of the prefix *regulation of* (e.g., GO:0042127 is REGULATION OF CELL PROLIFERATION and GO:0008283 is CELL PROLIFERATION). This observation predated the inclusion of REGULATES as a formal relation in the ontology. A similar analysis, which employed stemming and synonymy information, discovered interesting relationships linking terms in the Gene Ontology to terms in three other OBO ontologies [114].

A third connection between ontologies and text mining is that some approaches to biomedical text mining exploit ontological information to improve their performance. These approaches are premised on the idea that we can retrieve and extract information from documents more accurately if we take advantage of background knowledge about the domain of interest. Clearly, humans make extensive use of background knowledge when comprehending text, and thus the hypothesis that this approach will be beneficial for automated systems is compelling. In section 3.2, we discuss how PubMed indexes articles using MeSH terms in addition to the terms occurring in the title and abstract of the article. In section 4.3.3 we describe an information-extraction system that uses ontologies to determine the semantic roles of constituents in the sentences being analyzed.

---

3. For the remainder of this chapter, we use the term *ontologies* in a broad sense to refer to resources that encode domain knowledge, including controlled vocabularies, taxonomies, thesauri, and bona fide ontologies.

Although ontologies offer significant value for many types of biomedical studies, it is important to revisit a point that we made at the beginning of this chapter. Namely, that although there are many rich resources, such as databases and ontologies, representing biological knowledge in a structured format, much of the scientific community's knowledge is still represented only in unstructured text form. Thus, the existence of rich biomedical ontologies does not obviate the need for powerful text-mining methods.

2.6   Summary

In this chapter, we have discussed the primary sources of biomedical text, the challenging issues that arise in analyzing it, and some of the methods for preprocessing, representing, and eliciting meaning from it. Additionally, we have described ontologies and related concepts and discussed how ontologies are connected to biomedical text mining in several ways.

Several themes emerge from our discussion of natural language processing. First, many aspects of language are ambiguous. It is often the case that linguistic units and relationships have multiple possible interpretations, and thus a central challenge in NLP is to uncover the correct interpretation. Second, meaning can be represented at many levels in natural language, from subword components called morphemes to phrases that represent connections among distinct sentences. There are methods designed to represent and analyze text at each of these levels, and text-mining systems commonly employ a pipeline of such methods in order to extract meaning from multiple levels when addressing the task of interest. Third, many state-of-the-art NLP methods are based on models that are induced from labeled training corpora using supervised machine-learning methods. Fourth, it is often the case that NLP methods provide more accurate analyses of biomedical text when they are trained specifically using biomedical text.

In the subsequent chapters, we describe the predominant types of biomedical text-mining tasks and how systems for these tasks build on the methods and concepts discussed in this chapter.

3 Information Retrieval

In its most basic form, *information retrieval* is the task of finding a set of relevant documents in a large text collection. Naturally, the relevance of a document depends on our particular information need at a given moment. Most of us perform information retrieval on a daily basis, using search engines such as Google or, for searches specific to the biomedical domain, PubMed. The typical retrieval task performed using such search engines is known as *ad hoc* retrieval. Under this retrieval scenario, a user specifies a query, which is most often a Boolean combination of terms or words, and hopefully obtains all and only the documents within the database that satisfy the query conditions. Different users issue different queries while the same user may issue different queries at different times. During a short interactive session, even the same user may change his or her queries to express a refined or an altogether different information need.

It is often difficult to accurately express the information need using Boolean combinations alone. As experience shows us, when trying to express our needs as queries to search engines, relevant documents can be missed while irrelevant ones are retrieved. We will discuss the causes for these phenomena in section 3.2. To address the problem, another form of ad hoc retrieval is based on *similarity queries*. Under this framework, instead of forming an explicit Boolean combination of terms, a set of terms or words constitutes the query and is matched against the text collection using some similarity criteria. The documents that are most similar to the query are retrieved. We discuss this paradigm in detail in section 3.3.

In contrast to ad hoc retrieval, where queries are issued at any time an information need arises, another task in information retrieval is *text categorization*. In this case, the goal is to partition a set of documents into a number of categories, where the documents in each category share a topic of interest. Such categories may be a priori defined by experts—in which case we usually refer to the categorization task as *classification*. For instance, a collection of medical documents may be categorized based on the disease they discuss, where documents discussing lung cancer form one category and those that discuss leukemia form another. Alternatively, categories may be automatically uncovered or discovered by the categorization system itself, in which case we refer to the categorization process as *clustering*. These distinctions are further discussed later in the chapter. As each category is characterized by a topic, a category can be viewed as a collection of documents satisfying a certain query that defines the topic. Thus categorization turns out to be a retrieval task in which the set of queries characterizing the categories is fixed, and each document is tested against these queries to decide its category. Specific types of text categorization under this view are known as *routing* and *filtering*. We discuss text categorization and its subcomponents in section 3.5.

3.1 Example: The *BRCA1* Pathway (Revisited)

Let us revisit the example given in chapter 1 of searching for information about the *FANCF* protein, which is involved in the *BRCA1* pathway. Suppose we are specifically looking for information about *FANCF*'s interaction with other proteins.

To start, we perform a search using the Google search engine in the simplest manner. As Google's default Boolean operator is *AND*, we simply type the terms *FANCF BRCA1 interaction,* which is interpreted as a request to retrieve the documents containing all three words: *FANCF* and *BRCA1* and *interaction*. The retrieval produces a set of 4,370 hits,[1] covering websites and documents containing all three terms. Many of these documents discuss *BRCA1* interactions with other proteins or entities. For instance, one of them [65] states that *BRCA1* directly interacts with the protein *FANCA* – but not with other *FANC* proteins – and specifically *not* with *FANCF*. Clearly, looking through the 4,370 retrieved sites and references to actually find the proteins interacting with *FANCF* within the *BRCA1* pathway is still a difficult and time-consuming task.

To narrow our search, we can simply use a limited database rather than the whole web and issue the same Boolean query against the biomedical database PubMed. In this case, typing in the terms "*FANCF BRCA1*

---

1. Query results from www.google.com obtained on June 6, 2011.

interaction" results in just four hits.[2] All four references indeed discuss the required interaction among several *FANC* proteins, including *FANCF*, as a complex for activating the *FANCD2* protein. The latter directly interacts with *BRCA1*.

In this example, the Boolean query issued to PubMed resulted in a small and manageable set of articles that can be easily read to obtain the needed information. However, a small shift in our interest can quickly lead to a much larger set. For instance, issuing an almost identical query for the *FANCB* protein (rather than the original *FANCF*) results in 72 PubMed hits instead of four.

A researcher looking for proteins interacting with *FANCF* in the *BRCA1* pathway has another option. Rather than use search engines such as the general-purpose Google or the domain-specific PubMed, the researcher may search a database of interacting proteins, such as BIND [11, 16], DIP [49, 149, 270], IntAct [9, 104], MINT [33, 159, 282], or BioGRID [18, 25, 231]. All these databases rely on the literature to obtain information about protein interaction, typically employing curators who read through the text to identify the pertinent information. Text categorization can be used as a way to assist curators in finding the relevant articles and the relevant text within them. For instance, BIND and DIP employ text categorization to automatically obtain documents and sentences that are likely to contain information about interacting proteins. We revisit the application of text categorization for database curation in more detail in section 3.5.

Several common ideas, models, and methods underlie both ad hoc queries and categorization. All the computational tasks involved in information retrieval require the representation of documents, the representation of queries or of categories, and the means to check for relevance. The rest of this chapter introduces these areas and forms the basis for understanding how information retrieval methods can benefit biology and biomedical text mining.

3.2   Indexing, Keywords, and Boolean Queries

The fundamental structure underlying Boolean queries for ad hoc retrieval is the *index*. As discussed in Section 2.4, a fundamental step in text processing consists of breaking all the documents in the database

---

2. Query results from PubMed at www.ncbi.nlm.nih obtained on June 6, 2011.

into small units, called *tokens*. The tokens can be words, stemmed words, or short sequences of characters. Tokens are combined to form *terms*, which denote here a basic unit of text representation. Terms can be individual words, pairs of words (bigrams), grammatical or statistical phrases, and other possible constructs. An index is a data structure that contains all these terms (typically sorted alphabetically) and holds, for each term, references to all the documents in the database in which it occurs, as demonstrated in figure 3.1.

A Boolean query is executed by identifying the terms involved, searching for them in the index structure, identifying the sets of documents in which each term occurs, and executing a set operator (such as union or intersection) to retrieve the set of documents satisfying the query as a whole. For example, in figure 3.1, the query *blood pressure* will retrieve document 8, while the query *blood pressure OR blood* will retrieve documents 5, 6, and 8. Further information on the subjects of queries and indexing is available in books on information management, access, and retrieval [147, 267].

The set of terms included in the index typically corresponds to terms that explicitly occur in the documents. However, other terms correspond-

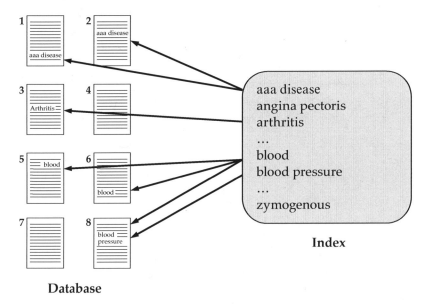

Figure 3.1
An index relating terms to the documents in which they occur.

ing to fundamental concepts may be associated with documents even if these terms do not explicitly occur in the text. For instance, each abstract in MEDLINE is assigned terms from the Medical Subject Headings (MeSH) [157]. MeSH is a controlled vocabulary, managed by the National Library of Medicine (as discussed in Section 2.5). The term assignment is conducted by human curators, who go through the full paper and decide which of the biomedical concepts from the MeSH list are relevant to it. MeSH terms thus become an integral part of the index structure, where each MeSH term is linked to the abstracts of papers to which it was assigned. Other examples of lists of terms that are being assigned to documents and, in turn, used as part of an index include the Unified Medical Language System (UMLS) meta-thesaurus [144], and the Gene Ontology (GO) [74, 246]. Although the GO is not used directly as a complete index structure, GO terms are used as a way to link genes and proteins with relevant publications. For instance, the UniProt database [257] associates proteins with publications that provide evidence for the process and function in which the proteins are involved. Processes and functions are denoted using GO terms, and references to the published literature that provide the evidence for the association between the protein and the GO term are given as part of the database entry.

An index structure can be further extended to include other types of "bookkeeping" information. Such information may include the exact positions within the document where the term occurs or parts of the text surrounding the term. This type of information supports identifying and locating longer phrases, whereas the explicit indexing only uses single terms. Other types of information can be grammatical, such as part-of-speech tags, or statistical, such as the number of times the term occurs within the document or the number of documents containing it. These additional types of information are used in practice to support certain processes and applications, including categorization, summarization, or information extraction from the retrieved documents.

The simple form of Boolean query, which has the advantage of efficient implementation over large databases, suffers several limitations for supporting domain-specific needs:

1. The number of documents typically satisfying short and simple Boolean queries is often too large to be of practical use.

2. A substantial portion of the retrieved documents are irrelevant to the user's information need.

3. Many relevant documents may not be retrieved.

Problem 2 above stems from the well-known polysemy and homonymy phenomena, while problem 3 stems from synonymy. Problems 2 and 3 are both are inherent to natural language understanding and were discussed in the previous chapter. One way to try and address these issues is to use a query mechanism that does not depend as strongly on explicit and specific words or terms. The vector model is an effective and widely used means to realize this idea, as described in the next section.

3.3   Similarity Queries and the Vector Model

A widely used alternative to the Boolean query is the *similarity query*, which is typically based on the *vector model*. In this setting, documents are viewed as vectors of weights, where weights are values reflecting the significance of the individual terms, as we formally define shortly. A query, q, is also viewed as a body of text rather than merely as a combination of search terms. As such, it may consist of many terms or even comprise a complete document. It is thus represented as a vector as well. The retrieval task reduces to searching the database for document vectors that are most similar to the query vector. A variety of similarity measures have been devised and used [79, 206, 226].

To formally define the vector model, we refer to the large set of documents from which retrieval is conducted as the *database,* and denote it as *DB*. The *vocabulary* of the database is the set of terms that occur within *DB*'s documents. Let M be the number of distinct terms, $\{t_1, \ldots, t_M\}$, in this vocabulary. As discussed before, a term, t_i, may be a single word or a longer phrase such as *blood pressure* or *acquired immunodeficiency syndrome*. Moreover, *stop-words*, which are non-content-bearing words such as prepositions and determiners, are typically removed in a pre-processing step.

A document, d, in the database is represented as an M-dimensional vector, $\langle w_{d_1}, w_{d_2}, \ldots, w_{d_m} \rangle$, where the weight, w_{d_i}, represents the occurrence or the significance of the term t_i within the document. The particular choice of term weights can greatly influence the results of a similarity search and several schemes are used in practice for calculating the weights. A few intuitive principles guide most weighting schemes [226]:

• *Local term frequency*   Terms that occur frequently within a single document are likely to be significant with respect to this document.

• *Inverse document frequency*   Terms that occur in many documents within the database are unlikely to be significant with respect to any particular document or query.

• *Inverse document length* A term that occurs frequently within a short document is more significant than a term occurring the same number of times within a longer document.

• *Relevance* Terms that are over-represented in documents that have been judged to be relevant for a specific query are likely to be significant with respect to that query.

The last criterion is based on the following intuition: terms that proved useful for identifying relevant documents in the past are likely to be useful in the future. We note though that it does have a form of cyclic reasoning: a term may indeed be viewed as significant if we already know that it is correlated with a relevant document, but before retrieval is performed, how can one know which documents are relevant? In accordance with the cyclic reasoning, this criterion is typically applied in an iterative process known as *relevance feedback*. Under this framework, a query is first issued by a user and results are returned by the system; the user then marks the documents he considers to be most relevant, and based on this information the system updates the weight vectors and produces new query results. This process is repeated until the user is satisfied with the results or until there is no further change in the weighting and in the document selection. A relaxed variant of this method that does not involve a user is known as *pseudo-relevance-feedback*; an initial weighting and retrieval scheme are used to produce results for a query, and terms are reweighted based on the documents that were ranked high in the retrieval results. The ranking of documents is based on their similarity to the query. Again, this process is iterated until there are no further changes in the weights or in the retrieval results.

We now discuss common weighting schemes in more detail. A simple and intuitive representation is binary weighting, in which a weight of either 1 or 0 is assigned to a term, corresponding to its presence or absence in the document:

$$w_{d_i} = \delta_{d_i} \overset{\text{def}}{=} \begin{cases} 1 & \text{if } t_i \in d, \\ 0 & \text{otherwise.} \end{cases} \tag{3.1}$$

Although this representation is straightforward, it does not account for any of the principles outlined previously, which take into consideration the distribution of terms within the documents and may improve retrieval quality.

A simple extension of the binary scheme uses the number of times the term actually occurs within the document as the weight (*local term frequency*). The intuition is that a document in which a query term is over-represented has a good chance to indeed be relevant to the information need represented by the query. Formally:

$$w_{d_i} \stackrel{\text{def}}{=} n_{d_i} \text{ if } t_i \text{ occurs } n_{d_i} \text{ times in document } d, \ 0 \le n_{d_i}. \tag{3.2}$$

This weighting scheme indeed accounts for the abundance of a term in a document but it does not consider the total length of the document. A short, concise, and highly relevant document may contain fewer occurrences of fundamental terms than a long, marginally relevant document. To reflect this, the above weight can be divided by the total number of term occurrences in the document, denoted by L_d (this is the *inverse document length* principle). That is, rather than use n_{d_i}, we use a normalized weight:

$$w_{d_i} \stackrel{\text{def}}{=} n_{d_i} / L_d. \tag{3.3}$$

This weight has a probabilistic interpretation, reflecting the probability of the term t_i to occur in document d; this interpretation is further discussed in section 3.4.1.

Yet another consideration is that if one query term frequently occurs in many documents and another is rare and specialized, documents containing the rare query term are likely to be more relevant to the user's information need than documents containing the frequent one (the *inverse document frequency* principle).

These principles are combined and formalized through a family of weighting schemes commonly known as $TF \times IDF$. The acronym stands for *term frequency $\times$ inverse document frequency*. Under this general scheme, the weight, w_{d_i}, is expressed as:

$$w_{d_i} \stackrel{\text{def}}{=} \begin{cases} r_{d_i} \cdot f_i & \text{if } t_i \text{ occurs in } d; \\ 0 & \text{otherwise,} \end{cases} \tag{3.4}$$

where r_{d_i} is a *local* measure of the occurrence of term t_i in document d, and f_i is a *global* measure, inversely proportional to the number of documents containing t_i in the whole database.

There are several ways to calculate the local measure r_{d_i}. We have already seen some examples, including $r_{d_i} = 1$ (equation 3.1) and $r_{d_i} = n_{d_i}$ (equation 3.2).[3] Other alternatives are $r_{d_i} = (1 + \ln[n_{d_i}])$ or

3. In both of these cases $f_i = 1$.

$$r_{d_i} = \left(k + (1-k) \cdot \frac{n_{d_i}}{\max_j [n_{dj}]} \right),$$

where k is a constant, $0 \leq k \leq 1$, and the denominator is the number of occurrences of the most frequent term in the document d.

Similarly, there are various options for calculating the global measure f_i. For example, denote by N_i the total number of documents containing the term t_i in the database DB. A simple expression for f_i is then:

$$f_i \stackrel{\text{def}}{=} \frac{1}{N_i}.$$

Other alternatives are

$$f_i \stackrel{\text{def}}{=} \ln \left[1 + \frac{|DB|}{N_i} \right],$$

where $|DB|$ denotes the number of documents in the database, or

$$\ln \left[\frac{|DB| - N_i}{N_i} \right].$$

Further discussion of weighting schemes is available in the extensive literature on information retrieval [147, 206, 226, 267]. A weighting scheme specific to biomedical literature was studied and presented by Wilbur and Yang [266].

Based on the vector-space representation, vector-similarity measures are used to assess similarity between pairs of documents or between a query and each document in the database. A measure widely used in information retrieval is the *cosine coefficient*. It denotes the cosine of the angles between the two vectors, as illustrated in figure 3.2. Formally, the cosine between two vectors, $\mathbf{V}^1, \mathbf{V}^2$, whose respective lengths (norms) are $\|\mathbf{V}^1\|, \|\mathbf{V}^2\|$, is defined as:

$$\cos(\mathbf{V}^1, \mathbf{V}^2) \stackrel{\text{def}}{=} \frac{\sum_{j=1}^{M} v_j^1 \cdot v_j^2}{\|\mathbf{V}^1\| \cdot \|\mathbf{V}^2\|},$$

where j ranges over the M vector elements.

3.4   Beyond Cosine-Based Similarity

Other approaches based on the vector-space model also aim to reduce the dependency of the retrieved documents on the explicit choice of

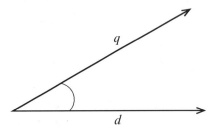

Figure 3.2
Determining the similarity between two vectors, q and d, using the cosine. The cosine of the angle between the vectors is 1 when the vectors perfectly coincide and is less than 1 otherwise. The similarity is 0 when the vectors are orthogonal, as would be the case for two documents that do not have any terms in common. Here we show the vectors in two dimensions but in a realistic information retrieval setting they would be M-dimensional, having one element for each term.

query terms, thus effectively improving retrieval. We discuss two such approaches: the first uses a probabilistic interpretation of the term vector as a sample from an n-dimensional distribution, while the second takes an algebraic perspective of the n-dimensional vector space and applies a decomposition operator to decouple the intended semantics from the explicit terms.

3.4.1 Probabilistic Language and Topic Models

A way to relax the dependency between retrieval results and the explicit query terms by using probabilistic models. Rather than require that a set (or a subset) of query terms occurs in a document, the retrieval task is viewed as that of finding documents that with *high probability* satisfy the need represented by the query.

To support this approach, probabilistic models have been devised to represent the query, the documents, and the user's information need. Van Rijsbergen's work is one of the earliest in this direction [259], and much of the pioneering research in this area is discussed in an extensive survey by Spärck Jones et al. [226].

Later work by Ponte and Croft [180] views each document in the database as a *language model*, which roughly corresponds to a multinomial distribution over terms. Simply put, when considering a vocabulary of M terms and a document d, one can count the number of times each term t_i occurs in d (denoted n_{d_i}), and divide this number by the total number of term occurrences in the document, L_d, thus obtaining a maximum likelihood estimate of the probability of term t_i to occur in the document d, $P(t_i \in d)$. The estimates for each of the terms can be used to construct an M-dimensional weight vector, $\mathbf{V}^d$ using the weights as defined in equation 3.3, $\mathbf{V}^d = \langle w_{d_1}, w_{d_2}, \ldots, w_{d_m} \rangle$, such that $w_{d_i} = (n_{d_i} / L_d) \approx P(t_i \in d)$. Using this weight calculation, the weight vector representing each document can be readily interpreted as a multinomial distribution over the terms, which is referred to as a language model.

A query is a set of terms and can therefore be viewed as a sample from an unknown language model. For every issued query, q, consisting of k terms $\{t_1{}^q, \ldots, t_k{}^q\}$, the probability of the query terms under each document's language model, $\mathbf{V}^d$, can be calculated as $P(q \mid d) = P(t_1{}^q, \ldots, t_k{}^q \mid \mathbf{V}^d)$. Documents d that give the highest values to $P(q|d)$ are viewed as the most likely to be the source language models for the query and are thus retrieved.

The idea underlying the language model was further developed by Berger and Lafferty [14]. They view a query as a result of a mental translation performed by the user. That is, they assume that the user's information need is ideally satisfied by a putative document. The user's query is essentially a translation of this ideal document into a short selection of specific terms. The retrieval task is thus posed as the identification of the documents in the database that are most probable to be the ideal source for which the query is the "translation."

A related approach was taken in the work on probabilistic latent semantics by Hofmann et al. [95], where documents are viewed as generated by a probabilistic model (similar to Ponte and Croft's language model) and the semantics of the chosen terms is stochastically determined by a set of hidden variables.

Another related approach, which was first applied in a biomedical context, is that of probabilistic themes [217, 213]. It uses a dual view to that of Ponte and Croft, treating the information need (i.e., the query, q) as a set of Bernoulli distributions—which is the source model, S_q. This model is learned via an iterative process of expectation maximization. Once the model is learned, the documents d that are most likely to have been generated by the source model, M_q, that is, those that give a high value to the probability $P(\mathbf{V}^d|M_q)$, are retrieved. This work is discussed further in section 6.3.2.

The idea of language and theme models has been further extended in work by Blei et al. [22] on topic models through latent Dirichlet allocation (LDA) to model mixtures of distributions corresponding to multiple themes or topics that together constitute the content of each document in the database.

A common idea underlies all the probabilistic methods described above. Namely, queries denote abstract concepts, regardless of the explicit terms that they use. The retrieved documents that best satisfy the query are those that carry the same abstract concepts. In the probabilistic framework, abstract concepts are represented as distributions over terms, where terms that are often used to denote the concept have a high probability, while terms that are unlikely to be associated with the concept receive a low probability.

Probability is not the only way to capture abstract concepts. Another approach, using an algebraic rather than a probabilistic method, is described next.

3.4.2   Latent Semantic Analysis

Latent semantic analysis is an algebraic approach to information retrieval that was introduced by Dumais et al. [51, 60, 61, 71]. It is based on two fundamental ideas:

• Abstract concepts are approximately conveyed by explicit terms. As discussed above, different term combinations may be used to identify the same concept (this is most commonly manifested through the use of synonyms). Moreover, the same term may be used to denote different concepts in different contexts (polysemy and homonymy). The *semantics* of a set of terms is the concept they convey. While words are overtly present in documents, semantics is not explicitly stated and is therefore latent. For instance, the concept of *lipid* could be described using terms such as *fat* or *fatty, sterol, wax,* and *glyceride* without explicitly using the term *lipid.* Thus the *lipid* concept may be latent in the text and conveyed only through the use of other terms that are explicitly mentioned in the overt description.

• A collection of documents, each represented by an M-dimensional weight vector, can be viewed as a matrix. As such, algebraic operators can be applied to it. One particular operator, namely *singular value decomposition*, is often used to identify and extract the most significant components of the matrix. These are its largest k singular values, where k is typically much smaller than the original number of terms, M. Each document in the matrix can thus be approximately represented as a linear combination of these k singular values, or equivalently, as a k-dimensional vector rather than as the original M-dimensional vector.

By combining these two ideas, each of the k largest singular values of a document collection is viewed as a surrogate for a class of terms with a common hidden semantics. Both queries and documents are transformed and expressed as vectors over these singular values rather than as vectors over M terms, and the similarity measure is applied to these transformed vectors, whose dimensionality is lower than that of the original term space.

This method is not often used in mainstream information retrieval because it has been shown effective only on relatively small collections of documents. Moreover, the algebraic transform to singular-values space hides the actual words in the documents. Thus, in its basic form the method does not provide the intuition or the ability to observe the terms responsible for the document similarity. However, further research on latent semantic analysis has been carried out within and outside the information retrieval community [110, 178]. In the biomedical domain, systems based on latent semantic analysis or on ideas directly related to it have been introduced in recent years [34, 99, 130], as discussed in chapter 6.

3.5   Text Categorization

Text categorization is the assignment of category labels, typically from a predefined set, to a text document. There are two main approaches to categorization. One is based on hand coding, in which a set of rules is manually defined to reflect expert knowledge of correct categorization. The main drawback of this approach is the knowledge-acquisition bottleneck. Rules must be manually defined by interviewing a domain expert and by accurately translating and encoding domain knowledge as simple operational steps. Any modification to the categories requires further intervention by a human expert. As such, this is not the method of choice when the number of documents to be categorized is large, the rules are hard to obtain and encode, or when domain experts are few and their time is scarce.

An alternative and common approach is based on machine learning [43, 62, 113, 132, 137, 138, 139, 140, 151, 152, 201, 209, 261, 276, 277]. A text classifier is viewed as a function that is learned by an inductive process. The text classifier learns to map documents to their class labels, using as a training set example documents that have already been correctly classified into a set of predefined categories. The resulting classification may be either *hard* or *soft*. Under hard classification a document is strictly assigned to a single category [112, 140]. By contrast, soft classification entails a ranking of the categories for each document, based on relevance. Under this approach, the classifier returns a number between 0 and 1 (called the *categorization status value* [CSV]), which represents the strength of evidence or the probability that the document belongs to a certain category. Documents can then be ranked with respect to each category according to their CSV. (See a study by Yang [275] for a discussion and further references.)

Machine learning makes an important distinction between *supervised* categorization, known as *classification,* and *unsupervised* categorization, known as *clustering.* In classification, as discussed previously, a set of class labels and a set of precategorized training examples are provided. In this case, the learning task is to create a classifier that correctly assigns class labels to yet-unseen documents. By contrast, clustering is the partitioning of examples into sets, without the provision of predefined labels or already categorized training examples. The goal here is to produce subsets (clusters) such that documents within a cluster are similar to each other according to some criterion, whereas documents contained in different clusters are dissimilar.

In practice, the most commonly used form of computational text categorization is based on supervised machine-learning methods. There is a wide variety of algorithms and paradigms used in this area. In this chapter we focus on two examples of text categorization that have been widely applied in biomedical text categorization: the *naïve Bayes* and *logistic regression* methods. These are discussed in detail below.

Another example worth noting is that of decision trees [24, 160, 189], which are used to express a cascade of tests, checking for properties that documents must satisfy in order to be placed under a certain category. The tree effectively encodes decision rules for determining the document's category. A very different and commonly used method, which inherently uses the vector model of documents, is based on *support vector machines* (SVMs) [261]. SVMs have been widely used in the context of text categorization since Joachims [113] first applied the method in this domain. Under the SVM paradigm, training examples from two categories are used to calculate a hyperplane that separates the two classes in a vector space (often of a higher dimension than that of the original data). As demonstrated in figure 3.3a there can be infinitely many separating hyperplanes, and the idea underlying SVMs is to construct the one that maximizes its distance from the training examples

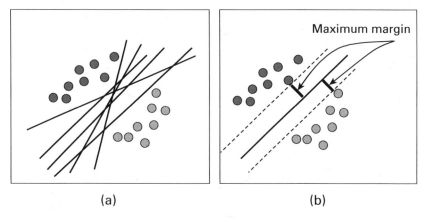

Figure 3.3
Classification hyperplanes in support vector machines. The dimensions of the space correspond to terms in the vector representation, and the circles represent instances of the two classes. (a) Some of the infinitely many hyperplanes (shown as lines in this two-dimensional space) that can separate the two sets. (b) The fundamental idea of support vector machines is to identify a hyperplane that is a maximum-margin separator.

closest to it, as shown in figure 3.3b. This idea is known as maximal margin separation. We now return to describe two additional classification methods in more detail.

3.5.1   Naïve Bayes

Naïve Bayes classification starts by representing each category as a probability distribution over terms. This distribution is estimated from the training examples. Once the model is built, the category selected for the document is the one that assigns the highest probability to the document's terms. We formalize this idea in the following paragraphs and provide an in-depth example. The method is known as *naïve* due to an assumption embedded in the computation and estimation procedures. Namely, given a category, the occurrence of each term in the document is conditionally independent of all other terms. We next formalize this notion.

Consider a document d, represented by its term weight vector $(w_1, \ldots, w_M)$. In the simplest case of naïve Bayes, we view each w_i as the value of a binary random variable, W_i, where $W_i = 1$ indicates the presence of the ith term and $W_i = 0$ indicates its absence. To classify the document into one of a set of possible categories, $\{c_1, \ldots, c_\kappa\}$, we are looking for the most probable category, which is the category that maximizes the conditional class probability given the document d. We use a random variable, C, to represent the unknown category, and compute the conditional probability $P(C = c_j|d)$ for each category c_j. Using Bayes' rule [52, 160], this probability[4] can be rewritten as:

$$P(c_j|d) = \frac{P(d|c_j) \cdot P(c_j)}{P(d)} = \frac{P(d|c_j) \cdot P(c_j)}{\sum\limits_{k=1}^{K} P(d|c_k) \cdot P(c_k)},$$

where the conditional probability $P(d|c_j)$ is the likelihood of document d to be generated as a sample from category c_j, $P(c_j)$ is the prior probability of documents to belong to category c_j, and $P(d)$ is the probability of document d to be generated regardless of any specific category. As can be seen, $P(d)$ is calculated as a sum over all possible categories $c_1, \dots, c_K$, where each term in the sum is similar in form to the product in the numerator. Thus, $P(d)$ is constant across all categories.

The fundamental assumption of naïve Bayes is the *naïve* assumption that the terms are conditionally independent of each other given the category. This assumption allows us to express the likelihood, $P(d|c_j)$, as a product of the individual term probabilities:

$$P(d \mid c_j) = P(w_1, \dots, w_M \mid c_j) = \prod_{i=1}^{M} P(W_i = w_i \mid c_j).$$

In some cases, we can also assume that the prior probabilities for all categories $P(c_1), \dots, P(c_K)$ are the same. Under this assumption, the category c_j that maximizes the conditional probability, $P(c_j|d)$, also maximizes the likelihood $P(d|c_j)$.

The individual probabilities, namely the probabilities of each term to occur in each category, $P(W_i = 1|c_j)$, are the model's parameters and are typically estimated from training data. We note that the probability of a term to not occur in one of the categories, $P(W_i = 0|c_j)$, is simply calculated as $1 - P(W_i = 1|c_j)$. The estimation process is demonstrated in the following example.

The curators of the Mouse Genome Database collect much information about the mouse as a model organism by scanning the published literature. Their initial scanning task is to decide whether a publication

---

4. When referring to the values that random variables take in probability expressions, we sometimes use a shorthand in which we write only the value, where the implied random variable is obvious. Thus $P(c_j|d)$ is a shorthand for $P(C = c_j|d)$.

is relevant for a particular subarea of the Mouse Genome Database or not. Our example task is to build a simple naïve Bayes classifier that, given a document represented as a binary vector of terms, classifies it as either relevant or irrelevant.

The classification task and the data we consider here are oversimplified to make the example easy to work through. Suppose we are given summary statistics from 20 relevant documents and 20 irrelevant documents. These statistics specify for each of three possible terms—*mouse, expression,* and *protein*—the number of documents containing them among the relevant and among the irrelevant examples, as shown in table 3.1. Using this training data, we can estimate the probability of terms to occur in documents for the relevant and the irrelevant categories, as summarized in table 3.2.

To conclude this example, we demonstrate the classification process itself. Suppose we are given a document, d, in which the terms *mouse* and *expression* occur and the term *protein* does not occur. Should this document be classified as *relevant* or as *irrelevant*? To answer this question we calculate the likelihood of the document d under the relevant and the irrelevant category models given in table 3.2.

$$
\begin{aligned}
P(d|C=rel) &= P(W_1=1, W_2=1, W_3=0|C=rel) \\
&\approx P(W_1=1|C=rel) \cdot P(W_2=1|C=rel) \cdot P(W_3=0|C=rel) \\
&= 1 \cdot 0.75 \cdot (1-0.2) = 0.6
\end{aligned}
$$

Table 3.1
Term occurrences in relevant and irrelevant documents for MGI curation

Term ID	Term	No. of relevant documents containing the term	No. of irrelevant documents containing the term
t_1	mouse	20	10
t_2	expression	15	8
t_3	protein	4	15

Table 3.2
Term probabilities for the *relevant (rel)* and *irrelevant (irrel)* categories

Term ID	Term	$P(W_i=1|C=rel)$	$P(W_i=1|C=irrel)$
t_1	mouse	20/20 = 1	10/20 = 0.5
t_2	expression	15/20 = 0.75	8/20 = 0.4
t_3	protein	4/20 = 0.2	15/20 = 0.75

Similarly, $P(d|C = irrel) = 0.5 \cdot 0.4 \cdot (1 - 0.75) = 0.05$. The likelihood of relevance (0.6) is greater than that of irrelevance (0.05), and (based on the assumption that the two categories are a priori equally likely) we categorize the document d as relevant.

In the above example, the probabilities of terms to occur (or not occur) in documents were estimated from the training data by counting term occurrences within the documents. However, in practice it is often the case that many of the terms rarely occur and therefore receive very low (or even zero) counts. Such extremely low probability values typically do not correctly reflect the true distribution of terms in all documents of interest. Rather, they are an artifact of a relatively small and possibly biased sample. Even more disconcerting is the fact that any term t_i that does not occur in the training set for a given class c_j would be assigned a probability of zero for this class, $P(W_i = 1|c_j) = 0$. If the term t_i occurs in any document to be classified, the classifier would automatically assign it a zero probability of belonging to class c_j, regardless of the document's actual contents. Moreover, extremely low probability values, when used as part of a computation, introduce underflows into the process. Therefore, it is standard practice to *smooth* the estimated model parameters [148], for instance, by adding small *pseudo-counts* to the actual counts used for estimating the probabilities.

In the above example, the model used for representing documents within a category c_j consists of multiple, independent Bernoulli distributions, $P(W_i|C = c_j)$, each representing the probability of a term t_i to occur in a document within class c_j. There is an alternative formulation of naïve Bayes for text classification in which a single M-dimensional multinomial distribution (rather than M independent Bernoulli distributions) represents the probability of terms to occur in each document. Under this formulation, a document is represented by a set of random variables, one for each token position in the document, where the multinomial distribution governs the probability of each random variable to be populated by any of the M terms in the vocabulary. A study comparing the Bernoulli and the multinomial models in the context of naïve Bayes classifiers for document categorization [151] suggests that the multinomial model can lead to better performance, especially when the vocabulary size is large. The multinomial model has also been used in much of the work on language and topic models in information retrieval [22, 95, 180], which were discussed in section 3.4.1. The multiple Bernoulli model, on the other hand, has been used as a basis for the probabilistic themes in the context of biomedical text mining [213, 217], as discussed in the same section and in section 6.3.2.

3.5.2 Logistic Regression

Another popular classifier that is often used in practice [208, 262] is based on *logistic regression*. Like naïve Bayes, this is a statistical method that fits a model to a set of training data and assigns the most probable category to a new data instance. However, the probabilistic model, the parameters used, and the underlying assumptions are all different from those of the naïve Bayes classifier.

Given a document d, represented by the vector $(w_1, \dots, w_M)$, a logistic regression classifier assigns d to the category c_j with the highest probability, $P(c_j|d)$, similar to the naïve Bayes classifier. However, logistic regression differs in several important respects. As shown above, the naïve Bayes approach uses Bayes' rule to compute the posterior probability $P(c_j|d)$ from $P(d|c_j)$ and $P(c_j)$, while calculating the likelihood $P(d|c_j)$ from its simpler model parameters, $P(w_i|c_j)$. By contrast, the logistic regression model directly calculates the posterior $P(c_j|d)$ from its parameters. To do this, the model consists of a set of parameters that can directly represent $P(c_j|d)$, as shown in the following. These parameters are estimated as part of the process of learning the classifier from the training data.

Logistic regression represents the posterior $P(c_j|d) = P(c_j|w_1, \dots, w_M)$ using the exponential form:

$$P\left(c_j|w_1,\dots,w_M\right) = \frac{e^{\left(\lambda_{j0} + \sum_{i=1}^{M}\lambda_{ji}\cdot w_i\right)}}{\sum_{k=1}^{K} e^{\left(\lambda_{k0} + \sum_{i=1}^{M}\lambda_{ki}\cdot w_i\right)}}, \tag{3.5}$$

where λ_{j0} assigns a weight for category c_j, and $\lambda_{j1}, \dots, \lambda_{jM}$ are parameters that play a role similar to the individual term probabilities, $P(w_i|c_j)$, in the naïve Bayes classifier. These parameters are learned from the training data, where each document d in the training set is assigned to its category c^d. The parameters are all estimated simultaneously to directly maximize the conditional probability of the category assigned to each document d in the training set, $P(c^d|d)$, over the whole training set. This is in contrast to naïve Bayes, in which the parameters are estimated independently for each class c_j.

Given a document to be classified, logistic regression calculates the linear function $\lambda_{j0} + \sum_{i=1}^{M} \lambda_{ji} \cdot w_i$ for each class c_j, and uses equation 3.5 to normalize the result into a probability value for each class.

The potential advantage of logistic regression over naïve Bayes is that the former often results in more accurate models because it is not as rigidly tied to the assumption that the terms are conditionally independent given the category, as explained in the estimation process. Which approach actually works better in practice depends on a combination of factors including, among others, the size of the training set and the size of the vocabulary [174].

A final word about terminology. The discussion of support vector machines as well as the example demonstrating the naïve Bayes classifier examined a special case of categorization: the case where there are only two classes, *relevant* and *irrelevant*. Text categorization for biomedical curation indeed often focuses on these two categories. For instance, the databases of interacting proteins [16, 18, 49, 104, 159] rely on identifying publications that discuss protein-protein interactions. The curators for these databases are interested in reading only interactions-related papers. All other papers are regarded as *irrelevant* for these databases. Similarly, curators for the Mouse Genome Informatics databases are interested in publications pertaining to mouse genes (*relevant*), while all other papers are regarded as irrelevant. The assignments of documents into one of two categories, relevant and irrelevant, in which the relevance criteria are fixed in advance, is often referred to in the literature as *document filtering*. If instead of a strict binary assignment into these two classes each document is assigned a relevance rank, the task is referred to as *document routing* [148].

3.6 Summary

In this chapter we have surveyed several approaches to information retrieval, which is the task of identifying text passages or documents that satisfy a user's information need. We started from the simple Boolean query model and worked our way through the idea of similarity queries and vector representations of documents to the broad domain of text categorization. The methods presented here and their many variations are discussed in detail in the literature on information retrieval [147, 206, 226, 259, 267] and on machine learning [160, 161, 204, 261]. Chapter 6 provides several concrete examples of the application of these methods within the biomedical domain.

4 Information Extraction

Chapter 3 focused on tasks that involve identifying which documents or passages in a large corpus are relevant to a given query. In some situations, however, one may want automated systems to perform a more fine-grained, in-depth analysis of the text. One type of analysis that may be useful is the identification of entities of interest, and of relations among entities. This is commonly referred to as the *information-extraction* (IE) task [32, 46].

Figure 4.1 provides an illustration of the information-extraction task. This example assumes that we are interested in recognizing entities of the types PROTEIN and LOCATION (where the latter refers to a compartment in a cell) and instances of the relation SUBCELLULAR-LOCALIZATION, characterizing in which compartments particular proteins are found.

The top of the figure shows two sentences in an abstract and the bottom shows the entities and the relation instance that should be extracted from the second sentence. This relation instance asserts that the protein *UBC6* is found in the subcellular compartment called the *endoplasmic reticulum.*

The information-extraction task of identifying instances of certain entity classes is called *named-entity recognition* (NER). The task of identifying instances of specified relations is often referred to as *relation extraction.* There is a wide variety of applications for which information-extraction methods are potentially valuable. For example, methods that perform relation extraction can be used to automatically populate databases. Given a database schema or some specified part of it, the IE system can process text sources to extract relation instances corresponding to the tables of interest. Using the terminology introduced in chapter 2, this application entails mapping from an unstructured representation (the text) to a structured representation (the database).

> "...
>
> *Here we report the identification of an integral membrane ubiquitin-conjugating enzyme. This enzyme, UBC6, localizes to the endoplasmic reticulum, with the catalytic domain facing the cytosol.*
> ..."

⇓

PROTEIN (*UBC6*)
LOCATION (*endoplasmic reticulum*)
LOCATION(*cytosol*)
SUBCELLULAR-LOCALIZATION (*UBC6, endoplasmic reticulum*)

Figure 4.1
An example of the information-extraction task. The top of the figure shows part of a document being processed. The bottom of the figure shows two extracted entities and one extracted relation instance.

Methods for named-entity recognition have an even broader range of applications. First, named-entity recognition plays a key role in relation extraction. In order to identify assertions about relations among entities, we must first (or simultaneously) detect which sequences of words refer to these entities.

Second, named-entity recognition is important for applications that index documents according to certain entity classes. Consider, for example, the iHop system [93, 94], which provides a browsable representation of networks of interactions among genes and proteins. (We discuss this system in more detail in chapter 6.) In order to assemble these networks, the iHop system must index articles according to the genes and the proteins discussed in them. It therefore must be able to recognize mentions of gene and protein names in the text.

A third, related use of NER methods is to identify and highlight passages of text that may be relevant to a particular information need. For example, some of the curators of the FlyBase database use a software system [117] that helps them scan the literature to find passages containing information that should be incorporated into FlyBase. One of the ways in which this system aids the curators is by highlighting putative gene names that have been identified using a named-entity recognizer.

The remainder of this chapter discusses in more depth the approaches that have been applied to the named-entity recognition and relation-extraction tasks. It also covers two closely related tasks, normalization and event extraction.

4.1 Named-Entity Recognition

Researchers and practitioners in natural language processing have been developing methods for the named-entity recognition task for more than twenty years [80]. The types of entities that have been considered extensively include PEOPLE, ORGANIZATIONS, and GEOGRAPHIC LOCATIONS, among others. In the field of biomedical text processing, the types of named entities that are of interest include GENES, PROTEINS, CELL LINES, CELL TYPES, CHROMOSOMAL LOCATIONS, DRUGS, and DISEASES AND DISORDERS.

There is a multiplicity of factors that make the task of recognizing biomedical named entities a difficult one. Consider, for example, the specific case of recognizing gene and protein names:

· There is great variety in the form that gene and protein names can take. Some names are short, individual terms, such as *BRCA1*. These are sometimes called gene *symbols*. For some species, the gene or protein names used in the literature almost always take this form. It is usually the case, however, that these short forms are merely abbreviations for longer, more descriptive names. *BRCA1,* for example, is short for *breast cancer type 1 susceptibility protein*. Depending on the gene and the species, such longer forms may be common in the literature.

· The names for some genes are homonyms of ordinary English words. For example, gene names and symbols in the fruit fly, *Drosophila melanogaster*, include *And, lot,* and *stuck*. Moreover, some gene names coincide with phrases that have common meanings in English (e.g., *cheap date*, *onion ring*, and *sunday driver*) or whimsical interpretations in English (e.g., *pray for elves* and *sonic hedgehog*). These cases suggest that a method for recognizing gene names should take into account the context of a candidate name and not just the properties of the candidate string itself.

· Because the same name is often used for a gene and for its products, it is sometimes not clear whether a particular mention refers to a gene, its corresponding protein product, its transcribed RNA, or perhaps some combination of these possibilities. This ambiguity is a complicating factor if the named-entity recognizer is intended to make distinctions among such alternatives.

· Some protein names are composed from other protein names. For example, *MAP kinase 1* is a protein name, as is *MAP kinase kinase 1*. Similarly, both *MAP kinase 8* and *MAP kinase 8 interacting protein 1* are names for distinct proteins.

Throughout this section, we center our discussion on the particular task of recognizing gene and protein names. We choose this specific task because (1) it is illustrative of the complexities involved in performing NER in the biomedical domain, (2) it has central importance to many biomedical text-mining applications, and (3) it is the biomedical NER problem that has been most widely studied. There are several publicly available corpora containing labeled gene/protein names that can be used for training and evaluating NER systems. Although there are cases in which we may want an NER system to distinguish between gene and protein names [84], we treat the two interchangeably in our discussion here. Many NER studies have not treated gene and protein names as separate classes because the same names and symbols are often used for both, and it is often not clear whether a specific occurrence of such a name is referring to the gene, the protein, or both.

4.1.1   Dictionary-Based Methods for Named-Entity Recognition

Perhaps the simplest approach to named-entity recognition is to look for matches against a dictionary of entity names [6, 31, 83]. A dictionary in this case is simply a list of names for each entity type. The NER task then becomes one of string matching. Given a passage of text to be processed, we can scan it for matches to names listed in the dictionary, checking to ensure that matches coincide with token boundaries.

There are two principal limitations of the dictionary-based approach to NER. First, it typically assumes that the provided dictionary is complete; that is, the method can recognize only entities that are listed in the dictionary. Consider the completeness issue in the context of recognizing gene names. Although there are databases that contain extensive gene-name listings for many model organisms, these databases cannot be considered complete because (1) the literature contains uncataloged typographical variants for some of the gene names indexed in the databases, (2) new gene names are still being coined for previously uncharacterized genes that are investigated in the lab, and (3) comprehensive databases do not exist for some organisms of interest.

A second limitation of the dictionary approach is that it cannot address the homonymy problem mentioned above (and discussed in chapter 2). Consider the extreme case of trying to recognize instances of the Drosophila gene name *And* using a dictionary approach. The approach is not able to distinguish occurrences of *And* that refer to the gene from those that refer to the common English word. In some cases, capitalization and typography may provide additional clues to help distinguish gene names from ordinary words. For example, if an occurrence of *And* is capitalized and appears in italics, then it is more likely to be a gene name. In general, however, additional context needs to be considered in order to resolve the homonymy problem.

The first limitation of the dictionary approach, incompleteness, can be somewhat alleviated by employing a *generalized dictionary*. The key idea underlying this approach is that many gene and protein names are composed of terms and symbols according to certain conventional patterns. A generalized dictionary consists of names that have been generalized to account for such patterns. By constructing a generalized dictionary, the approach can recognize previously unseen gene and protein names that are highly similar to those in the dictionary. This idea is illustrated in table 4.1, which shows some of the generalized dictionary names derived in one study by Bunescu et al. [31].

The leftmost column in table 4.1 shows a small sample of the 42,172 protein names that Bunescu et al. extracted from the Human Proteomics Initiative and Gene Ontology databases for their original dictionary. From this collection, they created a generalized dictionary by replacing certain terms with generic placeholders. Number terms are replaced by an <n> placeholder, terms consisting of single Roman letters are replaced by <r>, and Greek letters are replaced by <g>. For each original dictionary entry in table 4.1, the corresponding generalized dictionary entry is shown in the middle column. Bunescu et al. derived an even more generalized dictionary by stripping the generic placeholders from the entries in the original generalized dictionary. The resulting set of entries is referred to as the *canonical dictionary,* and is illustrated in the rightmost column of table 4.1. When using the generalized dictionary, a candidate protein name in a passage of text is identified as a protein

Table 4.1
Sample entries from generalized dictionaries of protein names used by Bunescu et al. [31]

Protein name	Generalized name	Canonical form
interleukin-1 beta	interleukin <n><g>	interleukin
interferon alpha-D	interferon <g><r>	interferon
NF-IL6-beta	NF IL <n><g>	NF IL
TR2	TR <n>	TR
NF-kappa B	NF <g><r>	NF

In the generalized names, <n> denotes a number token, <g> denotes a Greek letter, and <r> refers to a Roman letter.

name if its generalized form is an entry in the generalized dictionary. A similar procedure is used for matching against the canonical dictionary.

Not surprisingly, the generalized dictionary leads to higher levels of recall than the original dictionary, and the canonical dictionary provides even higher recall. However, the increased recall comes at the cost of decreased precision because the generalized entries lead to more false-positive predictions. As we discuss in section 4.1.3, these dictionaries do not have to be used in isolation but instead can be employed as one component within a named-entity recognizer that considers additional properties of each candidate name.

4.1.2   Rule-Based Methods for Named-Entity Recognition

As the generalized dictionary method illustrates, a useful source of evidence for recognizing named entities such as genes is the presence of common components in such names. These components do not necessarily have to be whole words, however. One of the earliest approaches for gene-name recognition applied a series of rules representing morphological and lexical properties of gene and protein names [69].

The approach, devised by Fukuda et al., consists of two stages. In the first stage, orthographic, morphological, and lexical rules are used to identify terms that may be part of protein names. In the second stage, lexical and part-of-speech analysis is used to identify sequences of terms that constitute protein names. This approach is illustrated in table 4.2.

Table 4.2
An example of applying the rule-based approach to protein name recognition

Stage 1:
The focal adhesion kinase (FAK) is ...
Ras guanine nucleotide exchange factor Sos is dependent upon ...

Stage 2:
The focal adhesion kinase (FAK) is ...
Ras guanine nucleotide exchange factor Sos is dependent upon ...

In the first stage, terms are identified using orthographic, morphological, and lexical rules. For example, *kinase* is tagged because it is a common component of protein names, and *FAK* is tagged because it is short and contains uppercase letters. Underlining is used to show the terms tagged in stage 1. In the second stage, terms are extended using rules based on lexical and part-of-speech properties. For example, the protein name seeded by *kinase* is extended to the left because it is separated from the preceding determiner (*the*) only by adjectives. The complete protein names identified in stage 2 are underlined.

The rules used in the first stage include the following:

1. Make a list of candidate terms. Include tokens such as *receptor* and *protein* that often form components of protein names. Include tokens that contain uppercase letters, digits, and nonalphanumeric characters.

2. Exclude tokens that are longer than nine characters and consist only of lowercase letters and hyphens (e.g., *full-length*).

3. Exclude tokens that indicate units (e.g., *aa, bp, nM*).

4. Exclude tokens that are composed mostly of nonalphanumeric characters (e.g., +/−).

The second stage of the approach involves two operations: merging and extending tokens identified in the first stage. The rules used in this stage include the following:

1. Merge terms identified in the previous stage that are adjacent to one another.

2. Merge nonadjacent terms that are separated only by nouns, adjectives, and numerals.

3. Extend terms to the left if there is a determiner or preposition and the intervening words are nouns or adjectives.

4. Extend terms to include a succeeding uppercase letter or Greek-letter token.

Steps 2 and 3 in this stage depend on using a POS tagger to determine the syntactic roles of the words being processed (as discussed in chapter 2). The terms that result from this stage of the pipeline are extracted as named entities.

The advantage of this approach is that the extraction procedure is easily understood by human experts and can be manually adjusted. Constructing and tuning such a set of rules typically involves a significant amount of human effort, however. Moreover, the handcrafted, rule-based approach has been largely supplanted by learning-based methods, which have demonstrated superior accuracy. We discuss such learning-based approaches in the next two sections.

4.1.3   Learning Classification Models for Named-Entity Recognition

As the examples in table 4.2 illustrate, there is value in considering orthographic, morphological, lexical, and syntactic evidence when trying to identify named entities. Instead of employing these types of features in hand-coded rules, more recent NER systems have used similar features in models automatically learned from labeled training data.

Whereas the hand-coding approach to constructing information-extraction systems requires a large amount of human effort to craft and tune extraction rules, the machine-learning approach may be able to automatically identify the regularities represented by such rules. The machine-learning approach, however, depends on having available a corpus of labeled training examples. Consider, for example, a corpus for learning an NER model. Such a corpus consists of passages of text along with annotations indicating which particular sequences of tokens in the passages correspond to instances of various entity types. These annotations are often referred to as *labels*.

In the simplest case, a set of possible labels may include only two types, say PROTEIN and OTHER. If the model was being trained to recognize additional entity types, then the set of labels would be extended to represent these other entities as well. Moreover, for a given entity type we may choose to have multiple labels. For example, if we believe that the properties of the first token in a protein name are somewhat different from the properties of successive tokens, then we may expand our label set to {BEGIN-PROTEIN, INTERNAL-PROTEIN, OTHER}, where the first label is intended to represent only the first token in a protein name and the second label is intended to represent subsequent tokens in a protein name. This is sometimes called a *B/I/O* (for *begin/internal/other*) representation. This idea can be generalized to include, for each entity type, separate labels for the last token in an entity and the special case of an entity consisting of only a single token.

A wide variety of machine-learning methods have been applied to the task of recognizing biomedical entities [225]. Let us consider one specific method in some detail. Bunescu and colleagues [31] empirically compared six different learning algorithms and a few other nonlearning methods on the task of recognizing protein names. Their experiments showed that the *maximum-entropy* learning method produced the most accurate NER models among those they considered.

In the maximum-entropy approach, a model is trained to approximate the conditional probability distribution $P(y|x)$, where x represents a short passage of text, such as a candidate protein name, and y denotes a possible label to be assigned to the passage x. In our discussion here, we will assume that x is a single token but the approach can be generalized to classify a passage of arbitrary size. A maximum-entropy model represents this distribution[1] using the following form:[2]

$$P(y \mid x) = \frac{exp\left(\sum_k \lambda_k \cdot f_k(x, y)\right)}{\sum_{y'} exp\left(\sum_k \lambda_k \cdot f_k(x, y')\right)}. \tag{4.1}$$

Here $f_k(x, y)$ represents a *feature* of the instance being considered, λ_k denotes a parameter that weights this feature, and y' ranges over all possible labels. Each feature $f_k(x, y)$ is a binary function based on the given token x and the candidate label y. For example, the kth feature might take the value 1 when the current token, x, is a noun and the candidate label, y, is PROTEIN, and take the value 0 otherwise. The parameter λ_k would then represent how predictive the *noun* POS tag is for the token having the PROTEIN label. The model would use similar "current word is a noun" features for the other possible labels, and each of these features would likely have a different weight associated with it in the trained model.

A machine-learning approach to NER, such as the maximum-entropy method, can employ a wide array of features, and through the training process, determine the extent to which various features are predictive of each label. Table 4.3 lists some of the types of features that have been used to recognize gene and protein names in learning-based NER systems.

As mentioned earlier in this section, the approach of constructing an NER system by hand-coding rules requires a large amount of manual effort. Given a labeled corpus for training, the learning-based approach

1. Maximum-entropy approaches are based on the principle that, subject to some constraints derived from the training data, the learned distribution should be as uniform as possible (i.e., have maximal entropy). The labeled training data is used to derive a set of constraints that specify class-specific expected values for features.

2. Note the similarity of the maximum-entropy model to the closely related logistic-regression model introduced in section 3.5.2. In the maximum-entropy model, the λ parameters are not class specific because the features are (i.e., each feature can take on a value of 1 only for a specific class). Thus, this maximum-entropy model can use a different set of features for each class.

Table 4.3
Some features that have been used in learned models for the biomedical NER task

Type	Example	Example matching token
word	word=*mitogen*?	*mitogen*
orthographic	is-alphanumeric?	*SH3*
	has-dash?	*interleukin-1*
shape	AA0	*SH3*
	A__aaaaa	*F-actin*
substring	suffix=*ase*?	*kinase*
lexical	is-amino-acid?	*Leucine*
	is-Greek-letter?	*alpha*
	is-Roman-numeral?	*II*
part-of-speech	is-noun?	*membrane*
dictionary	in-generalized-dictionary?	*interleukin-1 alpha*

The left column lists various types of features, the middle column lists specific instances of each type, and the right column lists tokens that match each instance. Shape features generalize tokens into terms represented using a four-character alphabet: *A* denotes an uppercase letter, *a* denotes a lowercase letter, *0* denotes a digit, and __ indicates any other character. Dictionary features are instantiated by testing tokens to see if they match entries in a given dictionary, as discussed in section 4.1.1.

requires some manual engineering as well. This effort, however, is invested primarily in defining potentially useful features rather than specifying exactly how such features can be used to recognize certain entity types. Moreover, the task of creating a labeled training corpus itself usually requires a considerable amount of manual work. There have been, however, investigations into various approaches for significantly reducing the amount of effort required to label a suitable training corpus [47, 163].

The task of learning an NER model, as described so far, is treated as a *classification* problem. The learned model is given some representation of a candidate entity (or part of one) and outputs a predicted class label for the candidate (e.g., PROTEIN or OTHER). A principal limitation of this approach is that it does not take into account any dependencies that may exist among the labels. Consider the case in which we are using the label set {BEGIN-PROTEIN, INTERNAL-PROTEIN, OTHER} for our NER task. With this label set, we need to ensure that we do not predict that an INTERNAL-PROTEIN immediately follows an OTHER label. Because INTERNAL-PROTEIN denotes a token within a protein name except for the first one, it should not be used to indicate the start of a protein name. Moreover, the classification approach to named-entity recognition does not take into account the expected length of protein names.

4.1.4   Learning Sequential Models for Information Extraction

An appealing alternative to treating the NER task as a classification problem is to employ a learning method that explicitly represents the sequential nature of the linguistic context in which names are found. Many of the most accurate named-entity recognizers have been based on probabilistic sequence models, such as hidden Markov models (HMMs) [44, 134, 220] and conditional random fields (CRFs) [100, 124, 142, 210]. These methods can readily represent dependencies among neighboring labels in a sequence and can take these dependencies into account when predicting labels for a given input sequence.

A hidden Markov model [148, 190] represents a joint probability distribution over sequences of tokens paired with states. In the simplest case, the set of states corresponds exactly to the set of labels. The HMM[3] represents the joint distribution as follows:

$$P(\mathbf{x},\mathbf{s}) = P(s_1 \mid s_0) \prod_{i=1}^{L} P(x_i \mid s_i) P(s_{i+1} \mid s_i).$$

Here, $\mathbf{x}$ is a sequence of tokens $<x_1, \dots , x_L>$ and $\mathbf{s}$ is a corresponding sequence of states $<s_1, \dots , s_L>$.

The factors $P(x_i|s_i)$ are called the *emission* probabilities and the factors $P(s_{i+1}|s_i)$ are called the *transition* probabilities. The emission probabilities represent how likely certain tokens are to be emitted, or explained by, particular states. The transition probabilities represent how likely each state is to follow any other state. The first transition probability $P(s_1|s_0)$ represents how likely it is to start a sentence in a particular state; s_1 indicates the state accounting for the first token in the sequence, and s_0 represents a special START state at which all paths through the model originate. The last transition probability $P(s_{L+1}|s_L)$ represents how likely it is to end a sentence in a particular state; s_L denotes the state accounting for the last token in the sequence, and s_{L+1} indicates a special end state where all paths terminate. The START and END states are *silent* in that they do not emit tokens. Figure 4.2 shows an HMM for recognizing protein names and illustrates how a path through the model corresponds to a labeling of a sequence.

---

3. Our description here focuses on a *first-order* HMM, in which each state s_{i+1} depends directly only on the previous state s_i. More generally, in an nth-order HMM, each state is directly dependent on the previous n states.

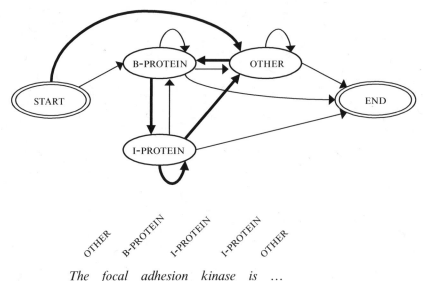

Figure 4.2
The state-transition structure of a simple HMM for recognizing protein names. The HMM has BEGIN-PROTEIN, INTERNAL-PROTEIN, and OTHER states in addition to START and END states, which do not emit tokens. The arrows indicate the allowable transitions (those with nonzero probability). Below the HMM is a fragment of a sentence that has been labeled for training. The darker arrows in the HMM show the path that should be used to explain this sentence.

The process of learning an HMM involves estimating these probabilities from training data, and in some cases, augmenting the set of states and the allowable transitions among them.

Given a trained HMM and a sentence, we can make named-entity predictions in the following manner: The sentence in this case is the sequence of tokens $\mathbf{x} \overset{def}{=} \langle x_1,\ldots,x_L \rangle$, and we would like to know the most probable sequence of labels $\mathbf{y} \overset{def}{=} \langle y_1,\ldots,y_L \rangle$ for $\mathbf{x}$. We assume that there is a mapping from states to labels. In a simple HMM, like the one shown in figure 4.2, we have one state per label. Therefore, each sequence of states $\mathbf{s} \overset{def}{=} \langle s_1,\ldots,s_L \rangle$ corresponds to a specific sequence of labels. Thus, we can determine the most probable sequence of labels by calculating $\mathrm{argmax}_{\mathbf{s}} P(\mathbf{s}|\mathbf{x})$.

Using the definition of conditional probability, $P(\mathbf{s}|\mathbf{x})$ for a given $\mathbf{x}$ and sequence of states $\mathbf{s}$ can be calculated using the emission and transition probabilities of the HMM:

$$P(\mathbf{s} \mid \mathbf{x}) = \frac{P(\mathbf{x}, \mathbf{s})}{\sum_{\mathbf{s}'} P(\mathbf{x}, \mathbf{s}')}.$$

Although the number of possible labelings is exponential in the length of the input sequence L, a dynamic-programming method called the *Viterbi algorithm* can be used to find the most probable of these labelings in $O(LS^2)$ time where S is the number of states.

The Viterbi algorithm operates by filling in a matrix $\mathbf{v}$ in which each element $\mathbf{v}(i, s)$ represents the likelihood of the most probable assignment of tokens to states that explain the first i tokens in $\mathbf{x}$ such that the ith token is aligned with state s. The matrix is filled in using the following recurrence relation:

$$\mathbf{v}(i,s) = \max_t \{\mathbf{v}(i-1,t)P(s \mid t)P(x_i \mid s)\}.$$

Here t ranges over the states that are possible predecessors to state s.

In addition to filling in each matrix value $\mathbf{v}(i, s)$, the algorithm also keeps track of which state t is chosen by the max operator. This information is recorded in a separate matrix of traceback pointers that we denote **ptr**:

$$\mathbf{ptr}(i,s) = \mathrm{argmax}_t \{\mathbf{v}(i-1,t)P(s \mid t)P(x_i \mid s)\}.$$

The final value calculated by the Viterbi algorithm is $\mathbf{v}(L, s_{\text{end}})$, which represents the likelihood of the most probable path that explains all L tokens of the given sentence and terminates in the END state. That is, this value represents the probability of the most likely labeling of the sentence among all possible labelings. The most probable path can be recovered by following the sequence of pointers from $\mathbf{ptr}(L, s_{\text{end}})$ back to $\mathbf{ptr}(0, s_{\text{start}})$.

As mentioned previously, the most compact HMM for a named-entity task will have exactly one state for each label. However, an HMM may have multiple states for some or all of its labels [221]. As long as there is a mapping from states to labels, we can calculate a predicted label sequence by calculating a predicted state sequence.

By identifying the most probable assignment of states, the Viterbi algorithm can be used to identify likely named entities. The state path calculated by the algorithm represents an alignment of sentence tokens to states, and those tokens that are aligned with states corresponding to named entities are extracted as entities.

The input sequences processed by an HMM can consist of more than a simple sequence of tokens. The tokens may be augmented, for example, by POS tags. In such a case, each $\mathbf{x}_i$ now denotes a vector of features $<x_{i1}, \ldots, x_{iF}>$ (the token, the POS tag, etc.), and the emission distributions must now represent the conditional probability $P(x_{i1}, \ldots, x_{iF}|s_i)$ where F is the number of features in the vector.

A limitation of the standard HMM approach is that, as we include additional features, it becomes more difficult to take into account the dependencies among them when representing $P(x_{i1}, \ldots, x_{iF}|s_i)$. Consider the features in table 4.3, which clearly have many dependencies among them. For example, the orthographic feature is-alphanumeric and the shape feature AA0 are strongly correlated. It would therefore require a complex model with many parameters to accurately represent $P(x_{i1}, \ldots, x_{iF}|s_i)$.

An alternative approach that gets around this limitation of HMMs yet still is able to take into account the sequential nature language is a type of model called a *conditional random field* (CRF) [131, 233]. Whereas an HMM represents the *joint* probability of sequences of tokens and states, $P(\mathbf{x}, \mathbf{s})$, a CRF directly represents the *conditional* distribution of states given the tokens, $P(\mathbf{s}|\mathbf{x})$. As we will see, this distinction allows us to use many features that describe the token sequence without having to specifically encode the dependencies among these features.

The representation used by a CRF for this conditional distribution is the following:

$$P(\mathbf{y} \mid \mathbf{x}) = \frac{1}{Z(\mathbf{x})} \exp\left(\sum_{i=1}^{L} \sum_{k=1}^{F} \lambda_k \cdot f_k(\mathbf{x}_i, y_{i-1}, y_i) \right), \tag{4.2}$$

where $f_k(\mathbf{x}_i, y_{i-1}, y_i)$ represents a particular feature referencing the observation sequence at position i and a particular sequence of states at positions $i - 1$ and i, λ_k denotes a coefficient that weights this feature, and $Z(\mathbf{x})$ is a normalization factor that ensures that $P(\mathbf{y}|\mathbf{x})$ is indeed a probability distribution by summing the feature weighting over all possible label sequences. The reference to the observation sequence at position i is denoted as a vector $\mathbf{x}_i$ because it should be understood as representing all aspects of the observation sequence $\mathbf{x}$ that are needed to compute features at position i. For example, one of the features might represent whether the word that comes right before position i is capitalized. Notice the similarity of this representation to the form of a maximum-entropy model shown in equation 4.1 (and logistic regression, discussed in chapter 3). The essential difference is that the features in a CRF, unlike those in a maximum-entropy classifier, allow dependencies between adjacent labels to be considered. We can think of each weight λ_k as encoding the compatibility of a particular combination of adjacent labels, y_{i-1} and y_i, with the observation sequence **x**.

Given an observation sequence and a conjectured label sequence, we can use equation 4.2 to assess how likely the label sequence is for the given observation sequence. As with an HMM, we can obtain a label sequence for a given input by finding the most probable such label sequence. This can be done efficiently using a slightly modified variant of the Viterbi algorithm.

The process of learning a CRF model requires using training data to optimize the λ parameters. Details on the particular optimization methods used can be found elsewhere [233]. Given a trained CRF model, predictions are made using essentially the same process as with an HMM. For an observed sequence, we can use the Viterbi algorithm to find the most probable sequence of states given the sequence.

The ability to incorporate many interdependent features makes CRFs well suited to problems such as named-entity recognition. As illustrated in table 4.3, biological NER typically includes a large number of features, with many dependencies among them.

4.2 Normalization of Named Entities

The task discussed in the previous section involves recognizing mentions of specified entity types in given passages of text. In this section we consider a related problem, sometimes called *normalization,* that entails associating each recognized mention with a canonical identifier for the entity being referenced. Consider the case of using named-entity recognizers to aid the curation of gene-related databases. In this task, we are given a set of canonical gene identifiers, such as the Entrez identifiers [171] for the genes in a particular organism, and the goal is to link each gene to the articles that discuss it. Simply recognizing gene mentions in articles is not sufficient for the task. Instead, we need to determine which mentions actually refer to each gene identifier on the given list.

Because normalization involves mapping predicted mentions to a closed set of names, one common approach is based on assembling a dictionary that associates each canonical identifier with all of its names and symbols. For example, if we were interested in recognizing references to human genes and mapping these mentions to Entrez identifiers, we might compile our dictionary from sources such as Entrez, UniProt [248], and the Hugo Gene Name Consortium [30]. Several factors make normalization a challenging task:

• Some names and symbols are polysemes for multiple genes. For example, there are more than twenty distinct *Drosophila* (fruit fly) genes that have the term *A* listed as an allowable synonym [90] in the FlyBase database. Similarly, there are many cases in which the same name is used for related genes in different species.

• Some gene names and symbols are homonyms for other, non-gene terms. Recall the examples of *And, lot,* and *stuck* mentioned in section 4.1.

• It is often the case that there are several different names for the same gene and a multitude of typographical variants for each one. The gene *tumor necrosis factor* is also referred to as *tumor necrosis factor alpha, TNF, TNF alpha, TNF-A,* and *TNF-α,* among other variants.

• Genes are sometimes described using paraphrases or permutations of standard names. For example, a gene with the name *ciliary neurotrophic factor receptor* may be referred to as *receptor for ciliary neurotrophic factor* [164].

To address these challenges, state-of-the-art approaches for normalization [146, 164] typically comprise two steps: (1) recognizing candidate gene mentions and (2) resolving ambiguous cases and filtering candidate mentions that are likely to be false positives. The first step can be handled using the approaches we discussed previously in this chapter. The second step commonly exploits additional sources of evidence to reduce ambiguity. Some of the ways in which various sources of evidence have been used for this second step include the following:

• References to species names in the context of gene mentions can help to resolve ambiguities that arise because the same gene name is used for genes in multiple species. There are taxonomies that can be used for this purpose [172], as well as specialized systems for species name recognition [75].

• A description of the gene occurring in the local context of a given mention can provide evidence useful for distinguishing which gene is being described. For example, to determine if a gene mention corresponds to a particular canonical identifier, Okazaki et al. [146] compute the cosine similarity (as described in section 3.3) between a vector describing the words found in the local context of the gene mention and another vector representing the description of the canonical gene identifier found in its Entrez Gene record [171].

• Other gene mentions in a given document can sometimes help to resolve ambiguous gene mentions [146]. For example, the introduction of a biological article will often specify the genes of interest, provide their full names, and also list abbreviations for those genes that are used subsequently in the article. An ambiguous mention, such as an abbreviation, that occurs later in the article may be disambiguated by using the name-to-abbreviation mapping that is provided in the introduction. Likewise, the concept of *discourse salience* can be used to disambiguate a given mention [48]. Discourse salience describes a situation in which certain salient entities are the focus of a document or passage. Salient entities are likely to be referred to repeatedly. Thus, if we can confidently resolve an entity mention to a particular identifier, we may conclude that this identifier is salient and thus a likely candidate identifier for other mentions in the same document.

4.3   Relation Extraction

To be able to automatically elicit rich biological knowledge from text sources, we must go beyond simply recognizing named entities; we must be able to identify relations between entities of interest. Among the relations that have been the focus of extraction systems to date are INTERACTION among pairs of proteins or other molecules, the SUBCELLULAR-LOCALIZATION of proteins, GENE-DISEASE associations, and TRANSPORT relationships that characterize how molecules are moved between cellular compartments by transporter proteins. Clearly there are many other relationships among biological entities that we may want to identify and extract, including generalizations and specializations of some of the above-mentioned relations. Iossifov et al. [105], for example, consider extracting various specializations of INTERACTION including direct relationships such as ACETYLATE, DEGRADE, and TRANSCRIBE, and indirect relationships such as ACTIVATE and SIGNAL. Most extraction systems have focused on extracting instances of binary relations (e.g., SUBCELLULAR-LOCALIZATION (*PRP20, nucleus*)), although a few investigations have considered relations with higher arity.

As with named-entity recognizers, relation-extraction systems have been devised by hand-coding patterns, via machine-learning methods, and by some combination of the two approaches. Likewise, relation-extraction systems developed to date have employed varying amounts of syntactic and semantic knowledge. Current state-of-the-art systems use detailed syntactic analysis and semantic categories and relationships [67, 70, 102, 105, 162], although simpler systems have been effective for some tasks.

4.3.1   Co-occurrence-Based Methods for Relation Extraction

One simple approach to relation extraction is based on finding co-occurrences, within short passages of text, of two entities that could be members of given relation of interest. For example, to find protein-protein interactions, we might search for abstracts (or even sentences) that mention two proteins together. Of course, co-occurrence does not necessarily imply that the relation of interest holds between a pair of entities. The text may assert that some other relation holds between the entities (e.g., it may state that two proteins are co-localized, not interacting) or it may not assert any relation at all. However, the precision of this general approach can be improved by considering other evidence in addition to the co-occurrence.

One such source of evidence is the frequency of co-occurrence relative to what is expected by chance. For example, Ramani et al. [192], who have extracted protein-protein interactions using a co-occurrence approach, assess the significance of the co-occurrence of two proteins, p_1 and p_2, as follows:

$$P(C \geq k \mid n,m,l) = 1 - \sum_{c=0}^{k-1} P(C = c \mid n,m,l),$$

where C is a random variable representing the number of co-occurring abstracts we might see by chance, k is the actual number of abstracts that mention both p_1 and p_2, l is the number of abstracts that mention p_1, m is the number of abstracts that mention p_2, and n is the total number of abstracts.

$P(C = c \mid n, m, l)$ is calculated using the hypergeometric distribution:

$$P(C = c \mid n,m,l) = \frac{\binom{l}{c}\binom{n-l}{m-c}}{\binom{n}{m}}.$$

In this way, we can determine if the number of co-occurrences exceeds what we would expect by chance, given the frequency with which each of the two proteins is mentioned in the corpus as a whole.

Another source of evidence that can be used to determine which co-occurrences are likely to represent protein interactions is the textual context in which the proteins are mentioned. Ramani et al., for example, further filter candidate interaction relations by applying a naïve Bayes classifier to the words in the abstracts containing co-occurrences. That is, their approach predicts that a pair of proteins interact if their co-occurrence frequency is unlikely to be attributed to chance and if the classifier indicates that their surrounding words are indicative of the language used to describe protein interactions. An early investigation into biomedical relation extraction used a similar approach to filter co-occurrences for the SUBCELLULAR-LOCALIZATION relation [47].

4.3.2   Rule-Based Methods for Relation Extraction

As with named-entity extraction, some of the earliest systems for extracting relations from the biomedical literature were based on hand-coded rules.

Table 4.4 shows some of the rules that were used by Blaschke et al. [20, 21] in an early system to extract protein-protein interactions. The rules in this system are regular expressions that are applied to sentences. The pre-processing that is done before applying these regular expressions involves first segmenting a given text into sentences and then segmenting each sentence into discrete tokens.

The rules in table 4.4 illustrate the breadth of patterns that can be characterized by regular expressions. Some of the rules incorporate specific wording, such as *"complex formed between"* in rule (iv). Others allow disjunctions of tokens, such as rule (iii), which allows the pattern to end with *complex, complexes, dimer,* or *heterodimer.* All of the rules

Table 4.4
Some regular-expression rules used by Blaschke et al. [20, 21] to extract protein-protein interactions

---

(i) **\<protein\>** \<word\>* \<verb\> \<word\>* **\<protein\>**
(ii) \<noun\> *between* \<word\>* **\<protein\>** \<word\>* *and* \<word\>* **\<protein\>**
(iii) **\<protein\>** \<word\>* **\<protein\>** \<word\>* [*complex/es, dimer, heterodimer*]
(iv) *complex formed between* \<word\>* **\<protein\>** \<word\>* *and* \<word\>* **\<protein\>**

Items in angle brackets indicate elements that can match a variety of tokens. The * character indicates that zero or more tokens may match the preceding element. Italicized items represent specific words to be matched. Bold items inside angle brackets represent the protein names to be extracted. Square brackets enclose sets of tokens, any one of which may match a given element.

allow variable-length sequences of unspecified tokens in various places, and some require words with a particular part of speech, such as a noun or verb, in places.

When a given sentence matches one of the rules, the two **<protein>** elements in the rule match particular proteins mentioned in the sentence, thereby identifying a potential instance of the INTERACTION relation.

Although proteins could be identified using a named-entity recognizer, Blaschke et al. simply used a dictionary of protein names in their work. Similarly, tokens matching the regular-expression elements denoted as *<noun>* and *<verb>* could be identified using a POS tagger. However, for this functionality, Blaschke et al. again used hand-constructed lists of nouns and verbs. The noun list contained words such as *activation* and *phosphorylation,* and the verb list contained words such as *activates* and *binds.* In subsequent sections, we consider relation-extraction approaches that take advantage of syntactic information that can be automatically predicted by current natural language processing methods.

4.3.3 Incorporating Syntax and Semantics into Relation Extraction

A limitation of the relation-extraction approaches we have considered up to this point is that they exploit very little information about the syntax and the semantics of the sentences being analyzed. Unlike the words that comprise a sentence, syntax and semantics are not directly observable. That is, sentences in most text sources do not come marked up with annotations indicating the syntactic and semantic roles of their various constituents. There are, however, computational methods and ontological resources that are able to assign syntactic and semantic aspects of sentences with fairly high accuracy.

State-of-the-art systems for relation extraction use rules or learned models that are based, in part, on syntactic and semantic analysis of the sentences being processed [67, 70, 102, 105, 162]. Consider the OpenDMAP system [102], which uses rules composed of words, phrases, parts of speech, syntactic structures, and semantic classes to identify relation instances in text. One task to which OpenDMAP has been applied is extracting information about protein-transport relations. Instances of this relation have at least three arguments: the protein that is transported, the location at which it originates, and the destination to which it is transported. An optional fourth argument, the transporting protein, can be extracted when it is mentioned in the text being analyzed.

Table 4.5
Some rules used by Hunter et al. [102] to extract PROTEIN-TRANSPORT relations

(i) PROTEIN-TRANSPORT :=	[TRANSPORTED-ENTITY] *translocation* (*from* {det}? [TRANSPORT-ORIGIN])? (to {det}? [TRANSPORT-DESTINATION])?;
(ii) PROTEIN-TRANSPORT :=	([TRANSPORTED-ENTITY] *translocation*) @ (*from* {det}? [TRANSPORT-ORIGIN]) @ (*to* {det}? [TRANSPORT-DESTINATION]);
(iii) PROTEIN-TRANSPORT :=	([TRANSPORTED-ENTITY dep:x] __ [action ACTION-TRANSPORT head:x]) @ (*from* {det}? [TRANSPORT-ORIGIN]) @ (to {det}? [TRANSPORT-DESTINATION]);

Elements enclosed in square brackets are the slots containing the arguments of extracted relations. The *?* suffix denotes elements that are optional, and the @ prefix indicates elements that are optional and can occur either before or after the required phrase. The variable x in rule (iii) indicates a syntactic dependency between the TRANSPORTED-ENTITY and the head of phrase specifying the ACTION-TRANSPORT.

OpenDMAP uses one or more rules to recognize instances of each defined *concept*. A concept may represent a semantic class or it may correspond to a relation. Table 4.5 shows several of the rules used by OpenD-MAP to extract instances of the protein-transport relation. When these rules match a given passage, three relation arguments get extracted: TRANSPORTED-ENTITY, TRANSPORT-ORIGIN, and TRANSPORT-DESTINATION. Rule (i) specifies that a PROTEIN-TRANSPORT instance may consist of a candidate TRANSPORTED-ENTITY followed by the word *translocation*, followed optionally by a phrase beginning with the word *from*, and an optional phrase beginning with the word *to*. After the word *from*, the first phrase optionally may have a word whose part of speech is a determiner and then has words matching the TRANSPORT-ORIGIN concept. The *to* phrase similarly consists of an optional determiner followed by words that match the TRANSPORT-DESTINATION concept. The concepts referenced in this rule, such as TRANSPORTED-ENTITY and TRANSPORT-ORIGIN, are themselves matched by rules. The concept TRANSPORT-ORIGIN, for example, is expected to be a cellular component and is thus associated with rules describing the names of cellular components listed in the Gene Ontology.

A rule may encapsulate a complex procedure for recognizing a concept. For example, OpenDMAP uses named-entity recognizers in LingPipe [251] and ABNER [210] to recognize protein names, which can instantiate the TRANSPORTED-ENTITY concept. One example of a phrase matched by rule (i) is "*.. Bax translocation to mitochondria ...*" where *Bax* is recognized as a protein by the named-entity recognizer ABNER.

Whereas the first rule specifies the order in which the key elements of a matching sentence must appear, rule (ii) relaxes these ordering assumptions. The @ symbol preceding the second and third elements of the rule indicates that these elements are optional and occur before or after the required phrase. Rule (ii) matches against the phrase "... *Bax translocation to mitochondria from the cytosol* ..." whereas rule (i) does not.

Rule (iii) extends the PROTEIN-TRANSPORT concept to include a slot that keeps track of the type of transportation action used in matching phrases. Moreover, this rule illustrates how additional syntactic knowledge can be incorporated into the extraction process. The wildcard symbol __ in this rule indicates that there can be some intervening tokens between the TRANSPORTED-ENTITY and ACTION-TRANSPORT elements. The use of the variable x in these elements, however, specifies that the TRANSPORTED-ENTITY must have a syntactic dependency on the head[4] of the phrase that contains the transport action. Thus the rule requires that, when there are intervening tokens between the TRANSPORTED-ENTITY and ACTION-TRANSPORT elements, there is evidence that the transport action refers specifically to the entity. This dependency would be determined by computing a syntactic parse of the sentence being analyzed.

4.3.4   Event Extraction

Although binary relations are able to represent many important biological facts, some relationships are not readily captured by binary relations or even relations of higher arity. Consider, for example, the sentence shown at the bottom of figure 4.3. This sentence describes how a protein-protein interaction (between *TRAF2* and *CD40*) is mediated by another relation. Specifically, when *TRAF2* is phosphorylated (by an unmentioned protein), this interaction is negatively affected. To capture the full meaning of this sentence, we need a representation that is more expressive than a single relation.

The top part of figure 4.3 depicts one structured representation that can be used to capture the relationships described in the sentence. This representation is based on the one used in the 2009 and 2011 BioNLP Shared Tasks on Event Extraction, which were community-wide evaluations of biomedical information-extraction systems [121, 122]. In this

---

4. The *head* of a phrase is the most important word in that phrase. In a noun phrase, the head is the main noun, in contrast to determiners and modifiers, which also may be part of the phrase.

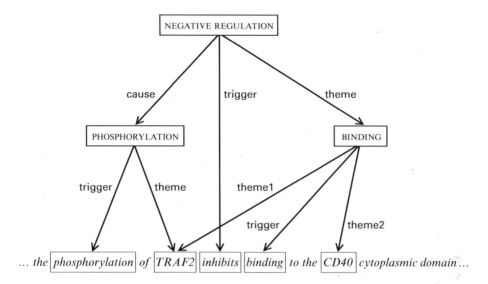

Figure 4.3
An example of the event extraction task. The bottom of the figure shows a sentence fragment being processed. The top of the figure shows the three inter-related events that should
be extracted from this sentence. Each event has several arguments indicated by the labeled
arrows.

representation, the content of the sentence is described as a set of linked
events. Each event has a type, a *trigger*, and additional arguments specifying the important constituents of the event. A trigger is a substring of
the sentence that is indicative of the event type. The principal argument
types are *themes*, which indicate the items (proteins or other events) that
are the subject of an event's action, and *causes*, which indicate the proteins or other events that are the cause of a given event. In the figure,
there is a PHOSPHORYLATION event whose theme argument is the protein
TRAF2, a BINDING event whose theme arguments are the proteins *TRAF2*
and *CD40*, and a NEGATIVE REGULATION event whose arguments are the
other two events. This event structure indicates that the PHOSPHORYLATION
event has a causal effect on the BINDING event; in particular, the former
negatively regulates the latter.

Through the BioNLP Shared Tasks, several groups have developed
systems that are capable of extracting descriptions of such inter-related
events [121, 122]. The approaches that have been developed for this task
employ many of the same types of NER and relation-extraction techniques that we have discussed previously in this chapter in addition to
classification methods as discussed in chapter 3. Many of the developed

systems use a pipeline approach in which a series of localized decisions is made in order to assemble events [19]. For example, some systems first try to recognize proteins and triggers, and then in subsequent steps try to link triggers to their associated themes and causes. More recently, several research groups have devised methods that are especially designed to take into account the dependencies among these various decisions, either in a pipelined approach [263] or in an alternative approach in which all of the relevant decisions are made jointly [154, 181, 200].

4.4   Summary

In this chapter, we described approaches for the two principal information-extraction tasks. The task of identifying instances of certain entity classes is called *named-entity recognition*. The task of identifying instances of specified relations among entities is referred to as *relation extraction*.

Two broad types of approaches have been used to develop information-extraction systems. One approach is to hand-code patterns, for either named-entity or relation extraction, using knowledge about the regularities in the way that entities and relations of interest are written. The alternative approach is to use machine-learning methods to automatically induce extraction models from labeled training examples. Currently, the most accurate approaches for named-entity recognition employ a machine-learning approach. For the relation-extraction task, however, both hand-coded and machine-learned methods are represented in the pool of state-of-the-art systems. It is important to keep in mind that the two approaches are not mutually exclusive. Further development of high-accuracy extraction systems will require exploiting both strategies.

Additionally, the most accurate information-extraction systems exploit a wide range of evidence in order to recognize entities and relations. The sources of evidence that are relevant include orthographic, morphological, lexical, syntactic, and semantic features.

Finally, we note that there a number of publicly available software packages that provide code for the tasks described here, including ABNER [210] and LingPipe [251] for named-entity recognition and OpenDMAP [102] for relation extraction.

5 Evaluation

From the discussion in previous chapters, it is clear that automated text mining and effective information retrieval can help realize a wide range of biological and medical goals. These goals vary in scope and domain; some examples of these goals, ordered in ascending level of difficulty, may include the following:

• Supporting curation of gene and protein information in organism-specific databases through focused, accurate retrieval;

• Providing easy access to information about bio-entities within displayed text by highlighting and hyperlinking such entities;

• Automatically reconstructing models of molecular networks from the published literature (which is an ambitious and not always well-defined task).

Generalizing beyond explicit words, sentences, or documents, text can be more broadly viewed as an additional, rich source of data complementing other forms of data such as gene and protein sequences, expression data, mutations and other genetic variations, and many others. We revisit some of these examples in detail in chapter 6.

While much work has been done in all these directions for over a decade now, a critical question is *how well* such systems perform. Every computational retrieval, extraction, and NLP system used in practice is judged by its performance, which is, in turn, evaluated either by popularity and user satisfaction or, more formally, through well-established performance measures. In this chapter, we focus on the latter—formal evaluation of text mining and retrieval systems. While measuring user satisfaction is a topic better covered by other disciplines such as human-computer interaction or marketing, we briefly touch on it here, primarily concerning its relationship to formal evaluation initiatives within the biomedical domain.

5.1 Performance Evaluation in Text Retrieval and Extraction

The evaluation of any system aims to answer a single question: *How good is it?* This question can easily be answered by yet another question: *Good for what?* The latter leads us to three main components, which are at the core of any text mining evaluation:

Task   A clear statement of what the evaluated system is supposed to achieve or do;

Gold Standard   A corpus of instances of the task at hand along with correct solutions assigned to the corpus;

Evaluation Metric   Objective functions by which one can quantitatively measure the performance of the system with respect to the task at hand. Such metrics are typically calculated based on the results produced by the system when applied to the gold standard.

Thus, an evaluation is carried out to quantitatively measure the merit of a text mining system with respect to the specific tasks that the system is designed to perform. If a system is supposed to retrieve text relevant to a user's needs, we should evaluate it based on how relevant the retrieved documents are to that same user's needs. It is important to note that quality can only be accurately evaluated with respect to specific, well-defined tasks. Thus, if there are multiple dimensions or parameters on which a system should be evaluated (such as ease of use, response time, or generalizability to other domains), all these dimensions need to be specified when defining the task and carefully taken into consideration when designing the gold standard and the evaluation metrics. We should not evaluate a system based on other criteria that were not specified at the onset.

Evaluation is typically carried out by comparing several systems that perform the same task on the same dataset. Specifically, for the comparison to be valid, the systems must run on the same text corpus. In order to know which systems are actually performing better, we must also know what the expected correct results are. Thus, as part of the task setting, we have a text corpus for which the correct expected answers were already given, and the systems are trying to replicate these correct answers. The text corpus along with the correct answers is typically referred to as the *gold standard* for the task. A common underlying assumption is that the gold standard data used in the evaluation is free of errors, although clearly such an ideal scenario rarely occurs in practice. Careful analysis of the results at the end of the evaluation is likely to help expose and address imperfections in the data. Moreover, realizing the fact that the data may be noisy and imperfect serves to motivate the development of more robust systems that can perform well even in the face of missing or noisy data.

To evaluate and compare different systems, we must also define the exact metrics by which the performance of each system on the gold standard is measured. Several objective functions corresponding to different aspects of performance are used in practice. We explore such objective functions in detail in section 5.2.

It is tempting to believe that by completing the evaluation and calculating the measures, one has obtained all the critical information about the relative performance of different systems, and can draw conclusions by "picking a winner." That is, it may seem reasonable that if system A performs better than system B within a controlled evaluation study, one can make a definite, general statement about the respective merit of system A compared to that of system B. For instance, when choosing a system, one may make a statement such as, *"Ernst and Rigor [ER03] reported that system A was superior to system B in their evaluations. We thus recommend using system A in our annotation pipeline."* Although such an inference may be valid in the ideal case, it does not typically hold true in practice.

There are several critical questions one must ask at the end of an evaluation experiment before drawing final conclusions about the practical utility of one system versus another. Important points for consideration include the following:

• *Significance of the results*   Are the reported differences in performance between the compared systems statistically significant? In particular, are the differences between the best-performing system and the runners-up statistically significant? Among all the compared systems, which pairs demonstrate statistically significant differences in performance?

• *Generalizability of the results*   Does the gold standard dataset provide a faithful, representative sample of the real world in which the system will be deployed in practice? Is it sufficiently large and complex to reflect reality?

• *Quality of the data*   Was the gold standard reliably constructed? That is, were the curators or annotators who provided the baseline of correct answers for the task consistent with each other? Were they reliable in terms of their relevant domain knowledge and their understanding of the task?

• *Generalizability of the task*   Does the evaluation task faithfully represent the actual goal for which the system is deployed? Does the task realistically reflect an actual users' needs? Is it general enough to cover a variety of future problems that the deployed system may end up addressing in practice?

• *Appropriateness of the evaluation measures*   Do the evaluation metrics correctly reflect the criteria by which the system will be judged in practice? Are there cut-off values that distinguish performance that is "good enough" from performance that would render the system unusable? Are these cut-off values based on realistic use cases?

For instance, consider a system whose task is to choose articles that may be relevant for curators populating the Mouse Genome Database. In practice, a human expert will go over the list produced by the system and manually select the articles that are actually relevant. In this case, the human expert may be willing to have a few irrelevant papers in the pile as long as no relevant ones are missed. The corresponding system can thus be imprecise and make false calls, as long as it does include in the list all the papers that are indeed relevant. Using the terminology defined in the next section, the preferred system in this case may be allowed to score low on *precision* as long as it scores high on *recall*. To summarize, an appropriate and reliable evaluation measure in this case should reflect the expected preference toward high recall along with the tolerance of low precision.

5.2   Evaluation Measures

This section presents some of the most common evaluation measures used in practice in the context of text retrieval, classification, and extraction. While the measures are described primarily from the point of view of evaluating a retrieval system, the same formulation and considerations hold true for evaluating classification and extraction systems.

5.2.1 Simple Measures of Retrieval

The most commonly used performance measures in both information retrieval and information extraction systems are *recall* and *precision* (see, for instance, [137, 267, 274]). These measures pertain primarily to tasks in which the goal is to distinguish between two types of instances (where instances can be named entities, terms, sentences, paragraphs, documents, or other forms of text): relevant versus irrelevant. We focus here mostly on such tasks because they are the most common in text mining applications. The measures of recall and precision are formally defined next.

Consider a set of n instances (where instances are as described above) that the system needs to label as "positive" or as "negative" according to some criterion. The criterion may be the relevance of a document to a specific query, the relevance of the document to a certain topic (category), or the membership of a term in a group of designated terms—such as gene names. Such labeling, by an imperfect system, partitions the original set into four subsets, as illustrated in table 5.1:

True positives   Instances *correctly* labeled as positive; we denote the number of such instances as TP.

False positives   Instances *incorrectly* labeled as positive; we denote the number of such instances as FP.

True negatives   Instances *correctly* labeled as negative; we denote the number of such instances as TN.

False negatives   Instances *incorrectly* labeled as negative; we denote the number of such instances as FN.

Table 5.1
The table illustrates the correspondence between system-assigned labels and the true labels in a dataset

		System-assigned Label	
		Positive	Negative
True Label	Positive	TP	FN
	Negative	FP	TN

Data instances are either positive or negative with respect to a predefined criterion. A system that attempts to assign the corresponding label to each instance may assign the correct labels—resulting in *True Positives* (TP) or *True Negatives* (TN), or incorrect labels—resulting in *False Positives* (FP) or *False Negatives* (FN).

The total number of instances in the set is $n = TP + FP + TN + FN$.

Precision, Pc, is defined as the proportion of true positives with respect to all instances that the system labeled as positive, whereas *Recall, Rc,* is defined as the proportion of true positives with respect to all instances that should have been labeled as positive:

$$Pc = \frac{TP}{TP + FP} \quad \text{and} \quad Rc = \frac{TP}{TP + FN}. \tag{5.1}$$

The concept of *recall* corresponds to that of *sensitivity*, which is often used within the broader context of the life sciences.[1] When designing a retrieval or an extraction system, there is typically a trade-off between precision and recall. We can think of two extreme cases. First, consider an "optimistic" system that labels everything as positive. For instance, a system that simply labels all documents as relevant, or a system that extracts all the terms occurring in a document and tags them all as gene names. Such a system indeed correctly identifies all the true positives and thus never errs by producing false negatives; however, it does give rise to many false positives. On the whole, such a system has a perfect recall, which comes at the price of low precision.

At the other extreme stands the ultimate "pessimistic" system, which correctly tags one positive example as positive, (e.g., accepts only a single truly relevant document as relevant, or tags only a single known entity as a member of the desired category). Such a system never makes a false positive error because it identifies only a single, truly relevant entity as positive. The recall associated with the system will thus be close to 0, while the precision is 1, because everything that the system has identified as positive is indeed correct. Both of these systems are trivial to implement and they each score very high on one performance measure, while being practically useless.

However, as pointed out in the previous section, in practice there are use-cases that may give preference to a system that demonstrates high recall with lower precision or vice versa. For instance, if one wanted to create a complete database of the literature published about a certain

---

1. Notably, the dual concept, *specificity*, which is usually calculated as $TN/(FP + TN)$, representing the ratio of correctly classified *negative* (or *irrelevant*) instances, is typically not used for evaluating retrieval and extraction. The reason for this is that typically most instances in a large collection are neither relevant nor retrieved. See further discussion of this issue in the context of the *accuracy* measure.

topic, one may prefer recall over precision in order to ensure the completeness of the gathered document set. In other cases, however, one may look for critical information when time is short and the stakes are high. In such a situation, obtaining results that are highly likely to be correct (high precision) is desirable even at the cost of possibly missing some relevant information (low recall). For instance, consider a physician in the emergency room, treating a patient with an acute and rare adverse drug reaction; the physician is looking for records of patients that have shown a similar adverse reaction but have been successfully treated by alternative means. In such a situation, the physician cannot afford to spend time reading through hundreds of records, trying to find the few relevant ones among them. Rather, the physician must receive just a few records in which the answer is present.

Two additional measures are commonly used to balance the trade-off between precision and recall, and estimate the overall performance of the system. One of these is *accuracy* [274], which simply evaluates the overall correctness of the system by calculating the ratio of correct answers with respect to the total number of answers the system has produced. Using the same notation as in equation 5.1, the accuracy is calculated as:

$$\text{Acc} = \frac{TP + TN}{TP + FP + TN + FN} = \frac{TP + TN}{n},$$

where n is the size of the whole dataset.

Accuracy is an effective measure for evaluating the performance for tasks in which there are more than two classes to consider (i.e., there are more than two possible labels for tagging the instances) and in which the classes carrying each label are balanced in size. That is, accuracy can be used to evaluate performance when the distribution of labels over the data instances is not heavily skewed toward any single label. However, this measure is usually not as useful for evaluating retrieval or extraction systems because in a large text database, most documents or instances are neither relevant nor retrieved. Thus, the true negatives *(TN)* dominate the numerator and inflate the accuracy measure without much regard to the relevance of the documents that were retrieved or, alternatively, to the correctness of the instances that were extracted.

The other measure, which combines precision, Pc, and recall, Rc, is known as the *F-score*. As it explicitly takes both precision and recall into account, it does not suffer the limitation of the accuracy measure in the face of a highly skewed data distribution. In its simple form [260], the F-score is expressed as:

$$F = \frac{2Pc \cdot Rc}{Pc + Rc}.$$

The F-score is a number between zero and one and it reaches one if and only if the system produces neither false positives nor false negatives. A more general form of the F-score, denoted F_β, allows the assignment of a higher weight to either precision or recall [218, 219, 274] and is calculated as:

$$F_\beta = \frac{(\beta^2 + 1)Pc \cdot Rc}{\beta^2 Pc + Rc}.$$

If the value of the β parameter is greater than one ($\beta > 1$), recall receives more of the weight, while $\beta < 1$ implies more weight is assigned to precision. When the parameter β is set to one, the same weight is given to precision as to recall, and the F_β measure reduces to the F-score.

A related measure, inversely proportional to a system's performance, is the *error score*, $E = 1 - F_\beta$. When using this measure, rather than maximizing a merit function, the objective becomes that of minimizing the error, E.

Finally, we note that the evaluation measures provided above are the basis for variations, where scaling and additional factors can be introduced to account for specific characteristics of the task at hand or of the end users' needs.

As mentioned previously in the discussion of accuracy, categorization tasks may involve more than just two classes. In such cases, the evaluation measures that are used in practice include precision (or specificity) and recall *per class*, where each individual class is considered separately as *relevant* and all other classes become *irrelevant*. The overall accuracy of the system can also be calculated, indicating the ratio of correctly classified instances when all classes are taken into account. One can also use a matrix representation to visualize the results, generalizing the one shown in table 5.1. In such a matrix, rows correspond to the true class labels and columns correspond to the assigned class labels. The entry in row i and column j contains the number of instances from the true class i that were assigned to class j. The numbers along the diagonal thus indicate the number of instances that were correctly classified. Such a matrix representation of classification results is known as the *confusion matrix,* and it is often helpful in summarizing results of multiclass categorization experiments.

5.2.2   Ranked Retrieval

As discussed in section 3.3, results from retrieval systems are often ranked lists of documents. When dealing with such *ranked retrieval* systems, one can look at a limited number of top-ranking documents and ask what the precision and the recall levels are for these documents only. Clearly, if we were to examine the whole retrieved list down to the lowest-ranking documents, the recall would be very high, while the precision would be low. Conversely, if we were to look only at the very top-ranking documents, the recall is likely to be low, while the precision would be high.

This idea is illustrated in figure 5.1. In the example shown, the database contains four relevant documents (shown in black within the figure). The system retrieves eight documents and ranks them according to its own relevance scoring. The top document is scored as most relevant and the bottom one is scored as least relevant. There are other irrelevant documents in the database, which are not retrieved by the system and are shown at the bottom of the figure. It can be seen that the recall changes at each point where another relevant document is retrieved.

If we were to return only the top-ranked document, ignoring all others, the precision would be 1, but the recall would only be 0.25. This recall level does not change if we were to return the two top-ranking documents because the second document from the top is irrelevant (shown in gray within the figure), although the precision drops to 0.5. However, if we return the top *three* documents in the list, the recall increases to 0.5 (two relevant documents are included in these top three) and the precision becomes 0.66. We can continue down the list, evaluating the precision at each point where the recall changes, and plot the *precision-recall curve*, as shown in figure 5.2.

Every rank in the list that is accompanied by a change in recall, which indicates a rank in which a truly relevant document is retrieved, is shown as a single recall-precision point in the plot. To compare different ranked-retrieval methods to each other, a comparison across whole precision-recall curves is needed. Comparing curves is not a simple matter because one retrieval method may dominate another at a certain retrieval rank but not at other ranks. An example of such a scenario is shown in figure 5.3. Two systems, system A and system B, are shown, where A demonstrates higher precision up to the point where 50% of the relevant

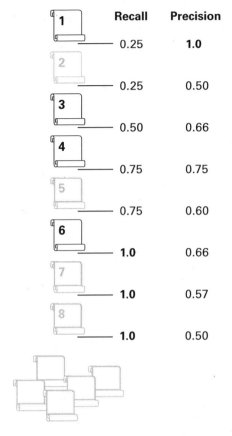

	Recall	Precision
1	0.25	1.0
2	0.25	0.50
3	0.50	0.66
4	0.75	0.75
5	0.75	0.60
6	1.0	0.66
7	1.0	0.57
8	1.0	0.50

Figure 5.1
A list of retrieved documents ranked according to the retrieval engine scores. Documents 1, 3, 4, and 6, shown in black, are relevant to the query. Documents 2, 5, 7, and 8, shown in gray, are irrelevant. The changing recall and precision are shown immediately below each document in the ranked list.

documents are retrieved; from that point on, system B shows a higher precision. In this case, the precision-recall curves do not themselves show which of the two systems is superior. It is important to note here, as discussed earlier in this section, that the regions of the curves on which we may wish to focus depend on the specific task and whether a higher precision or a higher recall is needed to satisfy the information need.

Several measures of retrieval evaluation are introduced for addressing the difficulty in assessing and comparing the quality of ranked lists. One such measure, known as *precision at n,* fixes a recall rank *n* and measures the precision at this rank. Another measure, *R-precision,* (where *R*

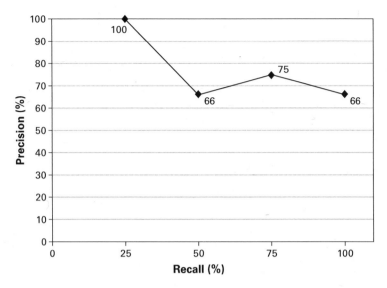

Figure 5.2
A precision-recall curve, showing for each level of recall (*X*-axis) the corresponding precision at that level (*Y*-axis). The precision values (in percents) are also noted on the curve itself.

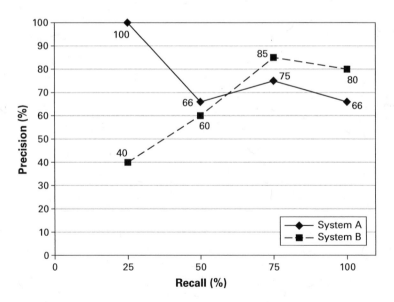

Figure 5.3
The precision-recall curves for two different retrieval systems. System A shows better precision up to recall level of 50%, but at a level of 75% and higher, system B demonstrates higher precision.

denotes here the number of relevant documents), measures precision at the point where the Rth document is retrieved, that is, precision at the point in which as many documents as are known to be relevant in the collection are retrieved. When evaluating ranked retrieval over a large number of queries, a measure often used by the Text Retrieval Conference (TREC) [243] is the *mean average precision* (MAP). It requires calculating the precision value at each rank where a relevant document is retrieved, averaging these precision values over all the considered ranks, and finally taking the mean of the averages over all the queries. Formally, for a set Q of queries and K ranks in which recall is measured, the mean average precision is calculated as:

$$\text{MAP} = \frac{\sum_{q \in Q} \left(\left(\sum_{r=1}^{K} (Pc(r,q)) \right) \big/ K \right)}{|Q|},$$

where $Pc(r, q)$ denotes the precision of a retrieval system at recall rank r when responding to query q. Additional, in-depth discussion of evaluation measures is available in the information retrieval literature [137, 265, 267, 274].

5.2.3   Using the Gold Standard Dataset

The data used in text-related evaluation efforts typically consists of a set of sentences, passages, or documents (which, in turn, may be abstracts or full-text articles) that were manually labeled by domain experts.[2]

For information retrieval and text categorization tasks, complete documents, passages, or abstracts are labeled as either *relevant* or *irrelevant*. For extraction tasks, specific occurrences of relevant entities or of relevant relationships are labeled individually within the sentences and passages in which they occur. The labeled instances may be considered as *positive* examples (while unlabeled occurrences are considered as *negative*) in the two-class case, or as examples of the various classes involved

---

2. While we refer to the people labeling the data as *domain experts*, they are often referred to as *human annotators*. Moreover, the labeling task is often referred to as *annotation*. However, the term *annotation* and its relationship to biomedical publications carry different meanings within biology (e.g., *genome annotation* refers to the task of labeling the genome with information about the genes; this information may be derived from text) as opposed to its meaning when used by computational linguists and NLP researchers (where the term refers to labeling the text itself with information, such part of speech tags). To avoid confusion, we do not use the term *annotation* here.

in the task if more than two classes are considered. The labeling is ideally done multiple times by several independent experts to ensure consistency and correctness of the data. Creating such a clean corpus is often labor intensive and time consuming, and as such, the available corpora are often relatively small.

As described in section 3.5, when discussing text categorization, the dataset is used for training and testing the systems. A critical component of a "fair" evaluation is that the data on which the system is trained should not be used for testing. The underlying idea is that a reliable system should be able to generalize beyond the data on which it was trained and make correct decisions on future cases. The separation of the training data from the test set can be achieved in two ways.

The first is to keep a subset of *held-out* data. That is, one simply reserves a portion of the labeled dataset for testing. The held-out portion typically consists of 10-50% of the labeled set; the actual ratio of training-to-test data typically depends on the size of the whole labeled dataset. In general, the larger the test set is, the more confident one can be about the estimated level of performance on future data. However, there is a clear downside to a large *held-out* test set; it comes at the cost of a smaller training set for the system.

Given the time and the effort that are involved in the careful manual labeling of a dataset, the labeled set is, at the onset, limited in both size and scope. By using, for instance, only 50% of the data for training, while leaving the other 50% for testing, we may effectively skew the evaluation by both training and testing only on subsets of the sample that do not fully represent the cases expected in practice. Specifically, we are testing a system whose training is not as thorough as it could have been had we used more of the labeled data, and we are testing it only on half of the available cases. However, if we use more of the labeled data for training, the test set may become too small to faithfully represent the expected performance of the system on yet-unseen cases.

A second way to get the most out of scarce data for training and testing, without obtaining any additional labels, is to use the data economically and effectively by "recycling" it. One way to do this, which is often used in practice, is known as *cross validation*. In cross validation, the data is partitioned into X equal parts (where X is typically five or ten). The system is trained on a dataset consisting of the union of $X - 1$ parts, (i.e., a fraction $(X - 1)/X$ of the whole dataset), and tested on the remaining, held-out part. The system is then retrained on a different subset of $X - 1$ parts, and tested on another single left-out part. This procedure is repeated

X times, resulting in X learning iterations and corresponding X learned models, in which a different subset of $1/X$ of the labeled set is used for testing while $(X - 1)/X$ of the set is used for training. In practice, it is common to use *fivefold* or *tenfold* cross validation, where $X = 5$ or $X = 10$, respectively, as well as leave-one-out cross validation. In the latter, $X = n$ where n is the number of instances in the labeled dataset.

Whereas the above paragraphs focused on ways to "make do" with scarce data by managing it judiciously, one can also try to obtain additional data for training in an efficient and cost effective manner. One such approach is to take advantage of unlabeled data as part of the training process. This can be done, for instance, through *semi-supervised learning*, where the training data includes both labeled and unlabeled instances [283]. Learning in this case often combines supervised methods (classification) with unsupervised ones (e.g., clustering). In a wide variety of applications, from document categorization to emotion prediction in spoken dialogs, semi-supervised methods have effectively boosted the performance of learned models by taking advantage of information derived from unlabeled instances. Another way to use unlabeled data is to employ an automated labeling process in a way that indeed provides more labeled data but possibly introduces some noise [47, 163]. That is, although some of the resulting labeled data may be mislabeled and may not be as clean as that labeled by human curators, the data can still be useful.

Another strategy for obtaining additional labeled data is through *active* learning [211]. The active learning paradigm is rooted in the idea that one can ask a human curator to label a few extra items, as long as one is selective with such requests. The goal thus becomes that of labeling examples that are most informative and would result in the most "bang for the buck." That is, a curator will be asked to label examples whose inclusion in the training set is likely to significantly improve the performance of the learned classifier. The field of *active* learning, which is based on early ideas about computer learning from queries and examples [7, 8, 77, 78], is concerned with automatically identifying data instances whose labeling by an expert will be most beneficial to the classifier's performance.

While the above methods are used to increase the amount of data or to make effective use of the dataset at hand, it has been shown in practice that relatively small datasets can still be effective for training a text classifier for use in the biomedical domain [54]. The specific application domain, the task, the literature involved, as well as the specific classification methods used and the selected features all have an impact on how effective the training is and how much data is needed for training a good-enough classifier for the task at hand.

5.3   Shared Evaluation Tasks

As discussed at the beginning of this chapter, a comparison of tools or methods requires an agreed-on task that is evaluated on a common reference corpus. For the evaluation to be useful in practice, both the task and the corpus of reference must be representative of the problems and of the data in the intended application area.

Several standard text collections as well as standardized retrieval and extraction tasks have been devised for supporting the development and the evaluation of text-processing systems. Two significant examples of standardized text collections that have been used in the past to evaluate text categorization systems are the Reuters collection [141, 199], and the OHSUMED collection of biomedical abstracts [88].

The Reuters collection is a set of about 800,000 news articles classified into thematic categories based on the focus of the news item. For instance, categories include MARKETS/MARKETING and FUNDING/CAPITAL, with subcategories such as BONDS, LOANS, CREDIT RATING, DOMESTIC MARKETS, or EXTERNAL MARKETS, respectively. Other tags assigned to articles in the Reuters collection include regional information and industry categories. The collection also provides standard tokenization of the corpus so that systems using the corpus can all utilize the same tokenization scheme. The Reuters collection has been used extensively for evaluating text classifiers, where the classification task is to label news articles with their correct category. The correct categories in this case are those assigned by the Reuters coders, either manually or through manual editing and correction of automated classification.

A different popular collection, directly related to biomedical text mining, is the OHSUMED corpus, developed at Oregon Health and Science University. This is a collection of approximately 350,000 medical abstracts, downloaded from MEDLINE during the years 1987–1991. The abstracts are tagged with the standard Medical Subject Headings (MeSH, discussed in section 2.5) as assigned by the curators at the National Library of Medicine. A subset of about 16,000 abstracts is also tagged with relevance judgments for a set of 106 queries posed by physicians. Each relevance judgment indicates whether a document is considered irrelevant, possibly relevant, or definitely relevant to the query. The OHSUMED corpus is often used for training and testing categorization systems aiming to automate MeSH assignment to documents and for evaluating information retrieval tools within the medical domain.

These two corpora are tightly coupled with the specific categorization tasks for which they were created and are not directly relevant for evaluating recent biomedical text-mining systems. To evaluate and encourage further development of text-mining tools specifically for biology, several specialized shared tasks have been organized in recent years. These include, among others, the Knowledge Discovery and Data Mining (KDD) Cup [119, 278], the TREC Genomics tracks [86, 87, 89, 203, 253], the BioCreative challenges [17, 92, 127, 135], the Elsevier Grand Challenge [63], the BioNLP shared task [121, 122], and the CoNLL-10 shared task [45].

5.3.1   The KDD Cup

The first such task, known as the KDD'02 Cup [119, 278], was organized as part of the International Conference on Knowledge Discovery and Data Mining (KDD) of the Association for Computing Machinery (ACM). The KDD'02 Cup defined a task pertaining to the curation process within the FlyBase database [245], which is concerned with genomic and proteomic information of the fly, *Drosophila*. FlyBase curators, when incorporating information into the database, rely on evidence elicited from the scientific literature published about *Drosophila*. Among other information, they look for reported evidence of experimental gene expression in the wild-type form (i.e., varieties that occur in nature without being modified in the lab) of the organism. For each *Drosophila* gene, the curators search for sentences and phrases within the literature indicating a new experiment that observed and measured a gene expression product. When a publication containing such phrases is found, its identifier is stored as part of the database entry for the gene, along with an indication whether the expression evidence reported within the paper pertains to a transcript or to a protein product.

One of the first two KDD challenges posed within the area of bioinformatics was to create an automated system that, given a set of full-text articles along with a list of gene names and synonyms (per paper), does the following:

• Identifies which publications contain experimental evidence of gene expression (for genes within a gene list provided along with each publication). That is, the system decides whether the paper should be curated by FlyBase curators.

• Specifies which of the genes mentioned in the article have their expression actually discussed in it.

• Indicates for each expressed gene whether the reported expression product is a transcript or a protein.

While human curators base their judgment on multiple sources of information, including their own experience as well as on the images within the articles, the data provided to the KDD Cup participants consisted only of the textual contents of the articles. The training set consisted of 862 full-text articles for which the specific gene information was provided by FlyBase curators. The test set consisted of 213 separate articles [278].

Thirty-two systems were submitted as entries to the KDD'02 Cup, and the F-score, calculated over the results produced by the different systems when applied to the 213 test articles, was used as the primary evaluation measure for comparing systems. The median F-score for the curation decision task was 0.58. For the task of identifying whether a gene was expressed or not expressed the median F-score was 0.35. The best-performing system was a rule-based information extraction system [197, 278], which obtained an F-score of 0.78 on the curate/do-not-curate decision and an F-score of 0.67 on the expressed/not-expressed decision. While there are obvious limitations to the evaluation process, such as the small test set, and apparent differences between the characteristics of the test set and those of the training set, this first evaluation task set the stage for several shared biomedical text mining tasks that followed.

5.3.2 TREC Genomics

A prominent and well-established forum for standardized evaluation of text retrieval systems is the Text Retrieval Conference (TREC) [243], sponsored by the US National Institutes of Standards and Technology (NIST) and by the US Defense Advanced Research Projects Agency (DARPA). TREC was formed in 1992 to support large-scale, comparative evaluation of retrieval systems. TREC is an annual event that offers multiple independent tracks, each of which focuses on different retrieval tasks from designated corpora, varying in size, in discourse domain, and in the information needs examined. Participants in TREC perform the task(s) designated for their track using their own developed systems. They submit their results for evaluation by judges selected for the task based on their domain expertise. Each task typically has its own specific evaluation measures, performance criteria, and its own team of evaluators.

The TREC Genomics track was formed in 2003, and focused on retrieval tasks pertaining to genomics, from full-text articles as well as from abstracts. The track was offered for five consecutive years [253]. Among the tasks defined in it were the following:

• The categorization of a set of about 0.5 million PubMed abstracts, based on whether or not an abstract discusses the function of a given gene, as well as the extraction of the passages summarizing the gene function from the abstract [86].

• The categorization of a set of full-text articles (about 20,000 publications) based on their relevance for curation by the Mouse Genome Informatics (MGI) databases [87, 89].

• Ad hoc retrieval of PubMed abstracts satisfying specific queries posed by physicians and biologists. A set of fifty queries reflecting biomedical information needs, along with a corpus of articles was given to the TREC participating teams. As an example, the TREC 2006 and 2007 corpus consists of about 160,000 full-text articles from 49 genomic-related journals. The teams submitted ranked lists of articles, retrieved by their systems for each information need. The ad hoc task formed an ongoing and evolving component at TREC genomics for several consecutive years [203, 253].

Describing and enumerating the extensive experiments and results involved in the TREC Genomics track are beyond the scope of this chapter. A clear conclusion from the track is that the performance achieved by current systems, as measured in the TREC Genomics evaluation, falls short of meeting the information needs posed as challenges in the task. For example, in terms of mean average precision (MAP), the best performing ad hoc retrieval systems reached a performance level of 0.33. This means that about 67% of the documents retrieved as relevant for a given information need even by the best performing system, are likely to be judged by human experts as irrelevant. The mean and median performance along the same measure achieved by all participating systems (calculated over 66 official runs) was below 0.20.

5.3.3 BioCreative

The BioCreative initiative takes its name from the aim of performing critical assessment of information extraction in biology [17, 92, 127, 135].

The event has been held every few years since 2004, and is past its fourth round.

The first BioCreative challenge introduced two main tasks: the first task concentrated on the extraction of gene and protein names from text and the second focused on curation of documents discussing gene or protein function, while identifying the evidence supporting the curation decision within these documents.

Task 1, the entity extraction task, consisted of two parts:

· Identifying and extracting gene/protein names in a set of sentences.

· Identifying gene/protein names in PubMed abstracts, and mapping each found gene/protein to its normalized canonical form. This type of mapping is typically referred to as *normalization,* as discussed in section 4.2. A list of canonical forms for the genes was provided with each abstract. Each gene reference found in the abstract also had to match the species discussed within the abstract (e.g., a reference to a human homolog, which is mentioned while discussing a mouse gene, should not be mapped to the canonical form of the mouse gene).

The datasets for the first part (gene/protein mention) of the task consisted of thousands of sentences for both training and testing, while the second part (normalization) used PubMed abstracts discussing mouse, fly and yeast genes.

For the name extraction subtask, 12 participating groups submitted 40 official entries. The highest precision, recall, and F-score are shown in table 5.2 along with the lowest results and the median [279].

For the gene/protein normalization part of this task, eight participants submitted 15 official entries. The highest and the median precision, recall, and F-score are shown in table 5.3 [90].

The best results in terms of precision, recall, and F-score were achieved for yeast. This can be explained by the fact that the gene nomenclature for yeast is relatively well-defined and uniformly used. The results for fly were overall somewhat lower, which again can be explained by the fact

Table 5.2
Results summary for BioCreative task 1A, gene/protein name extraction [279]

Measure	Highest	Median	Lowest
Precision	0.86	0.8	0.17
Recall	0.84	0.74	0.42
F-score	0.83	0.78	0.25

Table 5.3
Results summary for BioCreative task 1B, gene/protein normalization [90]

Measure	Highest (Y)	Median (Y)	Highest (F)	Median (F)	Highest (M)	Median (M)
Precision	0.969	0.94	0.831	0.659	0.828	0.765
Recall	0.962	0.848	0.841	0.732	0.898	0.730
F-score	0.921	0.858	0.815	0.661	0.791	0.738

Y denotes results for yeast, *F* for fly, and *M* for mouse.

that the fly gene nomenclature often includes common words (e.g., *armadillo, arm*) to denote fly genes. The results for mouse are similar to those of fly, with a better recall and a slightly lower F-score. A common problem with mouse genes is that their names are often similar or identical to those of homologous human genes, causing confusion in the creation of the gold standard and in the extraction itself.

The challenges presented in the second BioCreative [126, 127, 164, 225] still involved the gene identification and normalization tasks, as well as the extraction of protein-protein interactions from the literature. For training data, the BioCreative II Gene Identification task utilized the combined testing and training set from BioCreative I, while using 5000 new sentences for testing. The quality of all datasets in BioCreative II was improved upon with respect to BioCreative I by checking the data more rigorously for consistency in gene/protein name annotation prior to releasing the datasets to participants.

The task attracted a larger number of participating teams (21 participants), and the overall performance improved with respect to BioCreative I. The top-ranking team performed at a precision level of 0.885, recall of 0.859, and an overall F-score of 0.872. It is notable that four different teams achieved similar performance at these high levels, with no statistically significant differences among the top performers.

The gene normalization task focused on human genes (in contrast to the yeast, fly, and mouse genes that were the focus of BioCreative I), where the text was taken from PubMed abstracts rather than from the full text of articles. Twenty participating teams each submitted up to three runs to the competition. The gold standard for evaluation consisted of 262 abstracts annotated with a total of 785 gene identifiers. The top-performing system in terms of F-score exhibited an F-score of 0.810, recall of 0.833, and precision of 0.789. The top recall over all the systems was 0.875 and the top precision was 0.841.

The identification of protein-protein interactions in BioCreative II consisted of two subtasks. The first was to categorize full-text articles as relevant or irrelevant for protein-protein interaction curation, while the second was the actual extraction of the interactions from the text. The training data consisted of about 3,500 positive articles and about 2,000 negative articles. The test was conducted on a set of 338 positive and 339 negative articles. Nineteen teams participated in the classification task. The top F-score achieved was 0.78, with precision 0.703 and recall 0.876. Many of the teams achieved F-score of above 0.70 with recall at above 0.80. Participants used a variety of machine-learning classification methods such as support vector machines and naïve Bayes classifiers, as well as text-preprocessing methods taken from natural language processing including stemming, part-of-speech tagging, and shallow parsing. Performance on the extraction of interactions from the text was much lower, with precision, recall, and overall F-score all typically well below 0.40.

Although in-depth analysis of the results from these specific evaluation efforts is outside the scope of this book, one notable observation is that systems for identifying gene and protein mentions in the literature have matured during the past few years. Such systems, in their current level of performance, can be (and are) used as building blocks in support of other text mining tasks. Examples of such systems and tasks are provided in chapter 6.

Moreover, even when the performance of the participating systems at the current stage may not be as high as one may wish, an important component of the shared tasks is the creation of corpora and evaluation standards. The availability of the corpora allows for ongoing development and improvement of systems, thus having an impact long after the competition has run its course.

5.4 Summary

We have presented the main issues involved in evaluation of text retrieval and extraction systems, discussed several of the most commonly used performance measures, and provided examples of methods and datasets used for evaluating retrieval and extraction systems within the biomedical domain. While this book was being written, BioCreative II.5 and BioCreative III introduced several shared tasks, mostly focusing on protein-protein interactions. Other shared tasks and evaluation efforts, such as BioNLP, are ongoing as well.

Shared tasks often provide corpora of high quality and encourage researchers to work toward common, well-defined goals. However, these tasks, by their very nature, are focused only on a few problems and certain dimensions of text mining. Notably, we cannot consider the broad scope of biomedical text mining as synonymous with identifying specific types of entities, relations, or articles within the literature. Moreover, we should not consider any suboptimal results achieved by systems while competing on specific tasks over limited datasets as an indication that text cannot be effectively used to support biology and biomedical discovery. As demonstrated by several text-based systems that have been successfully applied — and are discussed in chapter 6 — text, accompanied by current methods for processing and mining it, already has much to offer in biology and medicine.

6 Putting It All Together: Current Applications and Future Directions

Throughout the previous chapters we have covered a variety of text-mining methods applicable to the broad range of tasks that are involved in obtaining information from text. In the beginning of chapter 1 we listed several goals within the biomedical domain that can be realized through the use of text. In this chapter we provide examples of systems and tools that have been developed to support such specific biomedical goals, and discuss in more detail the text-based methods that they employ.

6.1 Recognizing and Linking Bioentities

The identification of bioentities such as genes, proteins, small molecules, drugs, and diseases, as described in chapter 4, can support numerous tasks of interest including: (1) accessing the literature relevant to specific bio-entities, (2) identifying relationships among different entities, and (3) linking across a variety of data sources that provide further information about such entities.

A system that makes highly effective use of finding genes, proteins, and small molecules in text is the Reflect system [177, 195]. Reflect was the winning system in the Elsevier Grand Challenge 2008 [63], which aimed to develop and showcase systems that improve access to and use of scientific information published online in databases and journals.

Reflect uses a simple dictionary-based approach (as discussed in section 4.1.1) to identify genes and proteins in text. It provides a light-weight but important enhancement to web browsers, highlighting and hyperlinking in real-time genes, proteins, and small molecules mentioned within any text displayed by the browser.

The text in an HTML document is processed by the Reflect server before it is displayed. Organism names are recognized, and subsequently protein and gene names as well as identifiers of small molecules are found through matching against a comprehensive dictionary. The dictionary contains names and synonyms of the relevant entities and is constructed by compiling names from multiple public resources on genes, proteins, and chemicals, while allowing multiple orthographic variations.

The text is then displayed with the recognized named entities highlighted and linked to information that is shown through pop-ups when the entities are clicked on. Figure 6.1 illustrates the pop-ups and their links to the main web page. The information displayed for a selected gene or protein includes sequence and domain information (from the SMART database [136]), a graph showing significant interaction partners (from the STITCH database [232]), the best matching protein three-dimensional structure (according to the Protein Data Bank [183]), and information about the likely subcellular location of the protein. For small molecules, their two-dimensional structure is shown (from PubChem [184]), as well as significant interactions in which the molecules participate (from the STITCH database [232]).

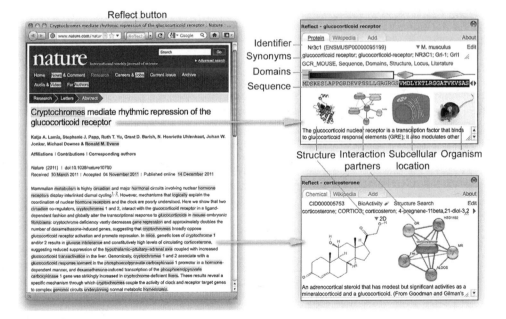

Figure 6.1
A screen shot illustrating the Reflect system. Genes, proteins, and small molecules are highlighted in the article shown in the browser window on the left. The pop-up windows on the right show detailed information about a selected protein (NR3C1) and a selected small molecule (corticosterone). (Courtesy of Sean O'Donoghue, December 2011).

Another readily accessible system that makes much use of both bio-entity identification and of linking entities throughout the biomedical literature is the iHOP web-based tool [93, 94, 103]. The iHOP system links genes and proteins with the scientific literature discussing them and detects relationships among genes and proteins through their co-occurrence in the literature. Unlike Reflect, which processes one web page at a time on demand, iHOP extracts its information by methodically pre-processing PubMed. The system scans through millions of PubMed abstracts and provides access to those that discuss genes and proteins. Given a gene or a protein name, iHOP displays the sentences containing the name, while highlighting and hyperlinking other entities that may be of relevance, such as genes, proteins, and MeSH terms (see section 2.5). Using a dictionary-based approach, genes and proteins are identified, as are MeSH terms, organisms, and verbs denoting potential interactions. The identified entities are displayed to the user in the context of the sentences containing them. Text denoting other genes and proteins is highlighted and hyperlinked to the sentences discussing these entities. Verbs denoting potential interactions are clearly shown. MeSH terms are also highlighted and hyperlinked to other sentences containing the same MeSH terms. A mechanism is provided for extending the query to Google or to PubMed using the gene/protein names and synonyms as well as MeSH terms. The iHOP system is uniquely extensive, as it associates tens of millions of sentences with over 100,000 genes and proteins mentioned in them, covering the literature for thousands of organisms.

An earlier system, built on the simple principle of identifying genes and proteins in text and linking them to one another based on their co-occurrence in the published literature, is PubGene [109, 185]. The system scans PubMed, searching for gene and proteins names, and graphically displays a network of genes and proteins that are related through co-occurrence. Nodes in the network represent genes or proteins, and connecting edges represent the co-occurrence of their names in the same PubMed abstract. The edges show the number of publications that support the association between the two connected entities. Figure 6.2 shows such a network as generated by PubGene. PubGene provides additional information about the genes and the proteins appearing in the networks, by linking them to Gene Ontology terms [74], MeSH terms [157], other chemicals and small molecules, as well as to their sequence data.

These examples demonstrate that relatively simple approaches, methods, and systems can have much value for end-user scientists by

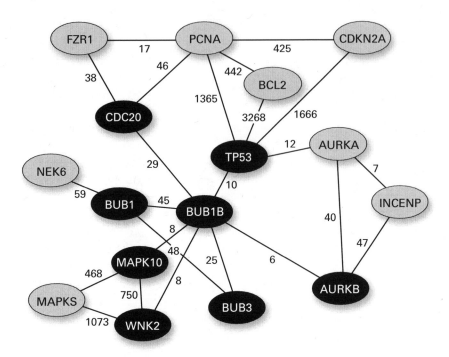

Figure 6.2
A screen shot of a network generated by PubGene. Nodes correspond to genes, and edges link together genes that are co-mentioned in the same PubMed abstract. The number of publications supporting each link is shown along the edge. Taken from the PubGene website, November 2010. (Image courtesy of Dr. Eivind Hovig.)

identifying and displaying entities (such as genes and proteins)—on a large scale— in the context of the text and by linking these entities to each other and to additional data sources. Moreover, the examples of iHOP and PubGene demonstrate how linking entities that occur within the literature, and visually displaying these links, can expose networks of relationships among genes, proteins, and other related entities and concepts.

6.2 Supporting Database Curation

Model organism databases, such as FlyBase [245], the Mouse Genome Informatics databases [158], the Saccharomyces Genome Database [205], and others [241, 255, 268], employ scientific curators who typically scan through the literature to find information (assertions and evidence supporting them) pertinent to genes and proteins in the specific organism.

Other public biological resources such as UniProt [257], JCVI's Comprehensive Microbial Resource [108, 179], and the enzyme database BRENDA [26] also employ curators to populate the database with high-quality information. The curators store the information and the supporting evidence along with literature references in the appropriate database entries. The work of curators can be simplified, expedited, and otherwise supported through automated tools for retrieval and categorization of relevant documents (as discussed in chapter 3), as well as through the extraction and identification of information of interest about genes and proteins (discussed in chapter 4). This section presents several examples of text-based tools and systems that have been developed especially to support curation efforts in biological databases.

One of the earliest database-specific search engines in the biomedical domain is Textpresso [37, 165, 166, 244, 258], which was originally developed by WormBase for the retrieval of *C. elegans* literature. Over the years it was extended to additional organisms and domains (including *Arabidopsis*, *Drosophila*, mouse, rat, zebrafish, neuroscience, and Alzheimer's disease, among others). For each domain or organism, Textpresso holds a set of tens of thousands of abstracts as well as full-text articles. Using extraction methods, it marks up and indexes entities of interest within the documents, such as alleles, processes, functions, and cellular components. By creating an index over the domain-specific entities of interest, Textpresso enables queries over these entities, providing accurate access to documents and to regions within documents in a way that is not currently supported by more general search engines such as PubMed. Further use of Textpresso as a basis for text mining in pharmacogenomics is discussed in section 6.3.3.

A different example of a database that uses text-mining methods for curating biological information is the comprehensive enzyme information database BRENDA [12, 26, 35]. BRENDA contains scientific information about all known enzymes, including kinetic, structural, stability, and genomic properties along with descriptors of methods for studying, isolating, and analyzing the enzyme. The information placed in BRENDA is based primarily on careful manual curation done by scientists trained in biology or chemistry. However, two accompanying databases, AMENDA (Automatic Mining of ENzyme DAta) and FRENDA (Full Reference ENzyme DAta), provide additional data, which are automatically extracted through text mining.

FRENDA links enzymes to PubMed references that provide information about the enzymes in specific organisms. In doing so, it recognizes synonyms of organism names and of enzyme names within the text. To support the association of publications in PubMed with enzymes and their organisms (under all their respective synonyms), FRENDA uses a dictionary of enzyme names and synonyms based on information that was originally curated in BRENDA, and a dictionary of organism names and synonyms based on the NCBI taxonomy of species names and phylogenetic lineages [172, 207]. Publications in which the organism and the enzyme co-occur are found using a Boolean search and are selected as candidate references for inclusion in the database. To reduce the number of irrelevant publications that are included, manually compiled lists of ambiguous enzyme and organism names are used to exclude matches that are likely to be spurious. FRENDA thus covers about 1.5 million PubMed references with over 400,000 organism-specific enzyme matches. These high numbers represent a strategy that favors high recall and lower precision (estimated at about 85% and 40%, respectively, according to a small study by the FRENDA team [35]).

AMENDA is a refinement of FRENDA that aims to associate each enzyme with an organism, source tissue, and subcellular localization. The latter two associations are supported through matching a candidate PubMed abstract in which an enzyme is discussed against tissue and localization dictionaries, where the tissue dictionary is based on the BRENDA Tissue Ontology [27], and the location terms are obtained from the Cellular Component part of the Gene Ontology [74]. Being more specific in its associations, and thus focusing on a smaller subset of abstracts, AMENDA aims at higher precision and lower recall than FRENDA. To improve the accuracy of association of enzymes with organisms, tissues, and subcellular locations while using co-occurrence of the respective terms in PubMed abstracts, AMENDA ranks the reliability of the associations it finds. For instance, the highest reliability of association between an organism and an enzyme is assigned when both the organism and the enzyme names occur in the title as well as in the same sentence within the abstract, and when the enzyme's EC number (Enzyme Commission number) appears in the PubMed abstract itself or among MeSH terms assigned to the abstract [35].

Textpresso, FRENDA, and AMENDA illustrate the applicability of text-mining methods in biological database curation. On the one hand, databases can be partly populated by automatic methods (such as those employed by FRENDA and AMENDA), and on the other hand, the automatic selection of text, as done in Textpresso, can be an initial step toward reducing the effort involved in manual curation by pointing out likely publications that the curators should explore first. These examples are by no means exhaustive. There is room both for the development of other ways to use textual data sources within the curation process, and for the application of existing methods to assist curation in other databases. Ongoing research and efforts on integrating text mining into the curation process include work by Karmanis et al. [118] on assisting FlyBase curators [245] by identifying genes and information about them in sentences through the application of natural language processing; work by Denroche et al. [54] on building classifiers to help identify publications containing experimental characterization of proteins in support of curation of the CHAR database [107]; and exploration of methods for assisting curators working for the Mouse Genome Informatics resource [158] in the Jackson Labs to identify relevant publications [87, 89] and to find pertinent information within these publications [58].

6.3 Text as Data: A Gateway to Discovery and Prediction

The two previous sections demonstrated the use of text as a source of information. In the systems we considered, retrieval methods are applied to identify publications containing useful information, and extraction methods are used to find specific entities and useful assertions about them within the text.

However, one can also think about text as a form of "raw data," whose analysis can reveal patterns and motifs leading to new discoveries. Similar to the way in which common subsequences shared across the genomes of multiple species, or homologies across proteins provide clues about underlying shared function, common recurring patterns in the text associated with biomedical entities can reveal hitherto unnoticed connections among these entities.

6.3.1 Discovery through Indirect Links in the Biomedical Literature

Early work by Swanson [234, 235, 236, 237] pioneered the use of text for the discovery of novel biomedical connections. Swanson observed that important but previously unestablished relationships can be revealed by following indirect, transitive links across concepts mentioned in the literature. A classical example of this approach is the case in which Swanson showed, by establishing links through the literature, that the use of fish oil is likely to be an effective treatment for Raynaud's syndrome [234]. As illustrated in figure 6.3, this discovery consisted of recognizing two

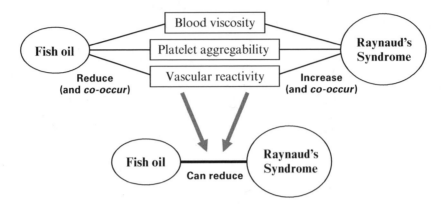

Figure 6.3
An indirect relationship between fish oil and Raynaud's syndrome (bottom part of the figure) uncovered by Swanson by linking two distinct sets of biomedical publications. The main two concepts—which were hitherto unrelated—are shown in circles. In the top part of the figure, the concepts in rectangles are the intermediate ones, whose relationships to the main concepts were already reported in the literature; the edges indicate co-occurrence of terms in publications and are labeled by the respective relationships as reported in the literature.

fundamental links within the literature: (1) reports about the ability of fish oil to reduce blood viscosity and decrease platelet aggregability, and (2) a separate set of publications that discuss increased blood viscosity and platelet aggregability as characteristics of Raynaud's syndrome. While the set of articles discussing fish oil's ability to reduce blood viscosity did not overlap with the set discussing increased blood viscosity in Raynaud's syndrome, Swanson was able to establish the yet-unknown connection that fish oil may be used for treating Raynaud's syndrome, through the common discussion of blood viscosity and platelet aggregability in the two publication sets. Two years after the publication of this relationship, an independent clinical study corroborated Swanson's literature-based hypothesis [55].

The fundamental idea presented here for linking concept *A* with concept *B* (where in the above example *A* is fish oil and *B* is Raynaud's syndrome) can be simplistically summarized as follows: Retrieve all the publications containing term *A* and all those containing term *B*. If these two sets of publications do not share a document, search for a set of concepts *C* that appear in both sets. The concepts *C* can indicate potential undiscovered relationships between concepts *A* and *B*.

While in the example *A* and *B* were both fixed a priori, the same idea can conceivably be applied toward open-ended discovery, where *A* and

B are not specified at the onset. However, the search space of all concepts and all biomedical publications in this case becomes prohibitively large. Thus, the basic idea has been extended and refined over the years by Swanson and colleagues [222, 238, 239, 240] and implemented in the Arrowsmith discovery system. Some of the extensions that have been investigated include: (1) generalizing A and B to be PubMed queries rather than prespecified concepts, (2) restricting the concepts in the set C to belong to specified categories, and (3) conducting searches in which only one of A or B is specified at the outset. Research along these lines has been further developed toward automated literature-based discovery by Weeber et al. [264], Wren et al. [269], and Srinivasan et al. [133, 227, 228].

6.3.2   Discovery through Thematic Analysis in Large-Scale Genomics

In the context of large-scale genomics and proteomics, methods have been introduced and developed to support biomedical discovery through the application of information retrieval and text categorization—both supervised (classification) and unsupervised (clustering). Specifically, technological advances in high-throughput gene expression analysis (e.g., the introduction of microarray technology) spurred research on the use of text from the scientific literature to interpret and explain commonalities within large sets of genes.

Early work in this direction by Shatkay et al. [213, 217] introduced a method for automated thematic analysis of text (similar to the approach more recently referred to as *topic modeling* [22]) in a way that can benefit the analysis of high-throughput biological studies, such as genomewide expression profiling. The method is based on two main ideas:

· The observation that text discussing a gene can be used as a surrogate representation for the gene.

· The observation that once genes are represented by their text documents and/or terms, they can be clustered using this text-based representation, thus exposing similarities among genes, and complementing other forms of gene-groupings derived from clustering gene sequences or gene-expression profiles.

Basing the method on information retrieval rather than on extraction of facts or gene names from sentences (see e.g., [20, 47, 109]) makes the method independent of gene nomenclature and of sentence structure, thus robust to the common problems of ambiguity typically associated with natural language processing, as discussed in chapter 2.

A database containing tens of thousands of PubMed abstracts pertaining to a specific domain of interest (for instance, all documents related to a specific organism such as yeast or all documents talking about AIDS) is used as a basis for discovery. To identify relationships among a large number of genes, each gene is first automatically mapped to a single publication discussing it in the database (e.g., an abstract curated for the gene by an organism database). This publication is then regarded as a *seed* representing the gene. A probabilistic theme model [217] is then learned, resulting in a set of 20 to 50 abstracts that are likely to be relevant to the seed abstract along with a set of terms summarizing the thematic contents of the document set.

An automated comparison of the abstract sets associated with the different genes is then used to derive relationships among genes, giving rise to groups of related genes. Again, the set of summary terms associated with each of the genes provides a putative, short explanation that justifies the grouping. Moreover, if the set of terms does not appear to provide a coherent explanation to the groupings, the discrepancy may serve as an indication that the grouping should be examined and not taken at face value.

For example, the genes in the resulting set may all be associated with terms such as *lipid, fatty,* and *fatty acid metabolism,* which can indicate that all the genes in the set are associated with the process of fatty acids or lipid metabolism. However, if some of the genes are associated with terms such as *fatty* and *lipid* while others are associated with *DNA replication* and *chromatin,* one may question grouping all these genes in the same set and go back to validate the clustering results. This verification ability stems directly from the use of text—whose semantics we understand—within the clustering process. By contrast, gene-clustering methods based on sequence or on expression profiles alone do not come readily equipped with clear semantics. Thus, clusters resulting from these methods may either be accepted as-is or need to be biologically validated through other means. An illustration summarizing the method is shown in figure 6.4.

The idea of establishing relationships among genes, as well as among other biological entities, based on similarity in the text discussing them has been employed by several other groups. Such work includes research on clustering genes and summarizing knowledge about them using associated text [229, 198], later publications by Homayouni et al. [99] and Chagoyen et al. [34] employing latent semantic analysis and non-negative

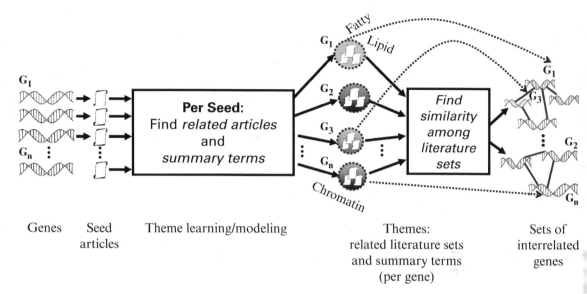

Figure 6.4
Connecting genes through shared themes: Genes (on left) are mapped to seed articles discussing them. A theme is built around each article, producing a set of related articles along with a set of summary terms per gene. Similarity among sets of articles is calculated, and genes associated with similar sets of articles are identified as related.

matrix decomposition, respectively, to group together genes whose literature profiles are similar; and work by Glenisson et al. [76] and by Küffner
et al. [130], which extended the above idea to group genes both by
expression profiles and by literature profiles, thus obtaining coherent,
interpretable clusters of co-expressed genes.

An approach closely related to the methods described above is the
characterization of gene sets by textual terms or by ontological categories that are over-represented in the annotations assigned to the genes.
As discussed in section 2.5, terms from the Gene Ontology have been
used to annotate the function of hundreds of thousands genes and gene
products. Given a set of genes (typically defined by selecting those genes
that exhibit some response of interest in a high-throughput experiment),
a system compiles all the ontology terms associated with the genes in the
set and selects as summary terms those ontology terms that are associated with significantly more genes within the set than expected by chance
(i.e., terms that are *enriched* or *overrepresented*). In particular, since the
inception of the Gene Ontology, a large number of tools based on this
approach were introduced to determine Gene Ontology terms that are

enriched for given gene sets [13, 59, 120], while newer tools based on similar methods are being used and developed [173].

6.3.3 Literature as a Source of Information for Large-Scale Gene-Disease-Drug Association

Current methods for genome-wide analysis of genes and disease aim to discover associations among genes, genetic mutations and drugs, as well as the potential for both therapeutic impact and adverse effects that drugs may have when interacting with certain mutations. As discussed in previous sections, retrieval and extraction tools can aid in methodological large-scale curation and literature can help expose new associations among biomedical entities. Thus, text-based tools hold promise to assist in large-scale gene-disease-drug association.

The field of *pharmacogenomics* is concerned with the large-scale study of connections between genes and drug responses, taking advantage of high-throughput methods [2]. The Pharmacogenetics Knowledge Base (PharmGKB) [123] is a comprehensive resource that integrates information from multiple sources about *pharmacogenes*, which are genes involved in modulating the response to drugs. PharmGKB stores information about these genes, important genetic variations in them, and drugs and their potential effects under regular conditions and in variants.

As with the curation efforts discussed in previous sections, the curators for PharmGKB scan through the literature and through existing relevant databases to associate known phenotypes with pharmacogenes and to obtain information about genetic variations and the functional implications such variations may have on drug response. PharmGKB curators also create summaries of the pharmacogenomics literature to compile a comprehensive list of well-characterized gene-drug interactions and phenotypes and to identify diagrams of the relevant drug-response pathways.

To facilitate PharmGKB's integration of literature and other sources of knowledge, the development of the database has been accompanied by the development of software tools for assisting curators and end-users in searching, finding, visualizing, and analyzing the diverse types of relevant data. In particular, these efforts involve devising automated tools to identify relevant publications and to extract pharmacogenomic information from them. To support these goals, PharmGKB is developing Pharmspresso [73], which takes advantage of the Textpresso suite of tools described in section 6.2, adapting it to identify pharmacogenomic-related full-text publications, and to find within them the information needed by

PharmGKB curators and users, such as references to human genes, polymorphisms, drugs, diseases, and their relationships. The results are presented as text segments in which the extracted concepts are visually highlighted.

6.3.4   Using Text Data as a Basis for Prediction

As noted at the beginning of this section, text can be viewed as a form of data; that is, words, tokens, or terms can be viewed simply as features, and statistical properties of text can be drawn from distributions over these features. (Sections 2.4 and 3.4.1 discussed how we can obtain such features and their statistical properties.) An important advantage of text-based features compared to other types of large-scale biological data, which was already mentioned in section 6.3.2, is their clear and readily understandable semantics. Unlike genomic and proteomic sequences or mass-spectrometry data, when looking at words, terms, or phrases, we typically grasp their meaning without the need for further interpretation.

Early work that employed the idea of incorporating text-based features into the analysis of biological sequences was a system developed by Chang et al. [36] to enhance homology search among protein sequences by simultaneously looking for similarity among the textual information associated with the proteins. In this approach, PSI-BLAST [3] is applied to the protein sequences themselves in order to find homology, while the text from the literature associated with the proteins through their entries in the SwissProt database [248] is tested for similarity using the cosine coefficient (discussed in section 3.3). While the results in this specific case were not shown to improve upon those obtained through sequence-based homology alone, the idea was novel and interesting and foreshadowed further research in related directions, as discussed below.

One significant biological task in which the use of text has been considered in several studies is the prediction of the subcellular locations of proteins. The subcellular location of a protein is the cellular component, or the *organelle*, in which the protein can be found within the cell. As the subcellular location of a protein can provide important cues about its function, its interaction partners, its role in processes and pathways, and its potential as a drug target, much research has been dedicated to the prediction of proteins' subcellular locations from sequence data [57].

The published literature can be used in two main ways in the context of learning more about protein localization:

1. As a resource from which known information about proteins' locations can be extracted using information-extraction methods [47, 194, 221];

2. As a source from which text-based features can be obtained to form a representation of a protein; this representation is then used for *predicting* the protein subcellular location.

Experiments employing the second idea were first reported by Stapley et al. [230] and by Nair and Rost [169]. However, the methods in these studies have not shown improvement over state-of-the-art, sequence-based prediction.

A newer set of systems, SherLoc [96, 215] and SherLoc2 [23, 28], introduced a new approach that improves the prediction of subcellular location for a wide variety of eukaryotic proteins through the use of both text and sequence data. The SherLoc system integrates several types of sequence features along with information derived from text. The PubMed abstracts associated with a protein are scanned as a source for text-based features, which are used to represent the protein. The system, which is based on support vector machine classifiers, uses articles curated by the SwissProt database as its primary text source, as was done by Chang et al. [36]. The system employs discriminative feature selection and integrates the text-based predictor with an accurate state-of-the-art, sequence-based location predictor, MultiLoc [97]. SherLoc was evaluated on standard benchmarks as well as on a new set of more recently localized proteins, devised specifically to test its predictive power, and demonstrated significantly improved performance compared to previously reported results, according to all standard evaluation measures (specificity, sensitivity, overall accuracy, and Matthew's correlation coefficient). By incorporating text into the location-prediction process, SherLoc measurably and significantly improved performance, thus being the first system to quantitatively demonstrate the utility of text as an important source of data in biological prediction.

Another example in which text is used as a form of data is an application concerned with finding similarities among bioassays in PubChem [82]. To identify relationships among bio-assays, their descriptions are taken from the PubChem database, converted into a vector representation, and clustered using the cosine-similarity measure. The method thus represents bioassays in a way that allows for a meaningful similarity measure to be developed.

In summary, we have described a variety of ways in which text is used as an additional rich source of data, complementing widely used forms of biological information such as gene and protein sequences, expression data, or genetic variation data.

6.4   Future Directions

This book has presented well-studied methods in text mining with a special emphasis on information retrieval and information extraction. Additionally, the book has described well-established evaluation procedures and demonstrated the applicability and the actual applications of text-mining methods in the biomedical domain.

Text mining within biology and medicine has been an active and fast-evolving field. Much progress has been made in the area of biomedical named-entity recognition and in the development of systems that support database curation and literature exploration. Other major advances are anticipated within the next few years. Current active research areas include the following:

• Data and text visualization in support of end-user needs. As shown previously in this chapter, one example of a recent system that pays much attention to the display of information in a way that supports both speed and ease of use is Reflect. The recent BioCreative challenge posed the Interactive Annotation Task, in which several interactive systems for supporting annotation were presented. The systems aimed to show textual information to curators in ways that would support and expedite their work. For instance, some systems explored techniques such as highlighting gene names, ranking genes by relevance, linking information to other genomic and proteomic resources, and allowing curators to add information to the contents already displayed (e.g., allowing the curator to add genes that she recognizes as important for curation within a displayed publication but that the automated process did not mark as such) [10, 15, 170, 202, 250].

• Integration of data and text from multiple sources, including electronic health records and the published literature. Medical text mining has been an active field within medical informatics for several decades, predating the sequencing of the human genome and the emergence of biological text mining as a field [40, 41, 66, 68, 106]. The work in this area has been motivated primarily by the need to identify pertinent textual information in patient records, clinical notes, radiology reports, and other clinically relevant sources, rather than by the abundance of publications in the biomedical literature.

As genome-wide assays of genotypes and molecular phenotypes of patients and pathogens are becoming prevalent, the boundaries between biological and medical information are becoming blurred and the integration of biological and medical data and text mining is a natural, rich, new, and promising research direction.

• Use of images within articles toward more effective text retrieval and mining. While the focus of this book was on mining the text in the biomedical literature, much of the information in scientific publications is conveyed visually through figures, graphs, microscopic images, photographs, and other types of imagery. Making effective use of the visual information within publications is an important new challenge that has only recently started to be explored [1, 38, 39, 53, 56, 85, 101, 125, 167, 168, 188, 191, 212, 271, 272, 280, 281].

References

1. Ahmed, A., E. Xing, W. Cohen, and R. F. Murphy. Structured correspondence topic models for mining captioned figures in biological literature. In *Proceedings of the Fifteenth ACM SIGKDD International Conference on Knowledge Discovery and Data Mining*. New York: ACM, 2009, 39–48.

2. Altman, R. B. 2007. PharmGKB: A logical home for knowledge relating genotype to drug response phenotype. *Nature Genetics* 39 (4): 426.

3. Altschul, S., T. Madden, A. Schaffer, J. Zhang, Z. Zhang, W. Miller, and D. Lipman. 1997. Gapped BLAST and PSI-BLAST: A new generation of protein database search programs. *Nucleic Acids Research* 25 (17): 3389–3402.

4. Ananiadou, S., B. Kell, and J. Tsujii. 2006. Text mining and its potential applications in systems biology. *Trends in Biotechnology* 24 (12): 571–579.

5. Ananiadou, S., and J. McNaught. 2006. *Text Mining for Biology and Biomedicine*. Norwood, MA: Artech House.

6. Ananiadou, S., D. Sullivan, W. Black, G.-A. Levow, J. Gillespie, C. Mao, S. Pyysalo, B. Kolluru, J. Tsujii, and B. Sobrai. 2010. Systematic association of genes to phenotypes by genome and literature mining. *PLoS ONE* 6 (3): e14780.

7. Angluin, D. 1981. A note on the number of queries needed to identify regular languages. *Information and Control* 51: 76–87.

8. Angluin, D. 1988. Queries and concept learning. *Machine Learning* 2: 319–342.

9. Aranda, B., P. Achuthan, Y. Alam-Faruque, I. Armean, A. Bridge, C. Derow, M. Feuermann, et al. 2010. The IntAct molecular interaction database in 2010. *Nucleic Acids Research* 38 (Database Issue): D525–D531.

10. Arighi, C. N., P. M. Roberts, S. Agarwal, S. Bhattacharya, G. Cesareni, A. Chatraryamontri, S. Clematide, P. Gaudet, M. G. Giglio, I. Harrow, E. Huala, M. Krallinger, U. Leser, D. Li, F. Liu, Z. Lu, L. J. Maltais, N. Okazaki, L. Perfetto, F. Rinaldi, R. Sætre, D. Salgado, P. Srinivasan, P. E. Thomas, L. Toldo, L. Hirschman, and C. H. Wu. 2011. BioCreative III interactive task: An overview. *BMC Bioinformatics* 12 (Suppl. 8): S4.

11. Bader, G., D. Betel, and C. Hogue. 2003. BIND: The biomolecular interaction network database. *Nucleic Acids Research* 31 (1): 248–250.

12. Barthelmes, J., C. Ebeling, A. Chang, I. Schomburg, and D. Schomburg. 2007. BRENDA, AMENDA and FRENDA: The enzyme information system in 2007. *Nucleic Acids Research* 35 (Database Issue): D511–D514.

13. Beissbarth, T., and T. P. Speed. 2004. GOstat: Find statistically overrepresented gene ontologies within a group of genes. *Bioinformatics* 20 (9): 1464–1465.

14. Berger, A., and J. Lafferty. Information retrieval as statistical translation. In *Proceedings of the Twenty-Second ACM SIGIR Conference on Research and Development in Information Retrieval*. New York: ACM, 1999, 222–229.

15. Bhattacharya, S., A. K. Sehgal, and P. Srinivasan. Online gene indexing and retrieval for BioCreative III at the University of Iowa. In *Proceedings of the BioCreative III Workshop,* 2010, 52–54.

16. BIND. Biomolecular Interaction Network Database. http://www.bind.ca, 2011.

17. BioCreative Critical Assessment of Information Extraction in Biology. http://www.biocreative.org/, 2011.

18. BioGRID. http://thebiogrid.org/, 2011.

19. Björne, J., F. Ginter, S. Pyysalo, J. Tsujii, and T. Salakoski. Scaling up biomedical event extraction to the entire PubMed. In *Proceedings of the 2010 Workshop on Biomedical Natural Language Processing.* Stroudsburg, PA: Association for Computational Linguistics, 2010.

20. Blaschke, C., M. Andrade, C. Ouzounis, and A. Valencia. Automatic extraction of biological information from scientific text: Protein-protein interactions. In *Proceedings of the Seventh International Conference on Intelligent Systems for Molecular Biology.* Menlo Park, CA: AAAI Press, 1999, 60–67.

21. Blaschke, C., and A. Valencia. 2001. Can bibliographic pointers for known biological data be found automatically? Protein interactions as a case study. *Comparative and Functional Genomics* 2: 196–206.

22. Blei, D. M., A. Y. Ng, M. I. Jordan, and J. Lafferty. 2003. Latent Dirichlet allocation. *Journal of Machine Learning Research* 3: 993–1022.

23. Brady, S., and H. Shatkay. EpiLoc: A (working) text-based system for predicting protein subcellular location. In *Proceedings of the Pacific Symposium on Biocomputing.* Hackensack, NJ: World Scientific, 2008, 604–615.

24. Breiman, L., J. Friedman, R. Olshen, and C. Stone. 1984. *Classification and Regression Trees.* Monterey, CA: Wadsworth and Brooks.

25. Breitkreutz, B., C. Stark, T. Reguly, L. Boucher, A. Breitkreutz, M. Livstone, R. Oughtred, et al. 2008. The BioGRID interaction database: 2008 update. *Nucleic Acids Research* (Database Issue) 36: 637–640.

26. BRENDA. http://www.brenda-enzymes.org/, 2011.

27. BRENDA Tissue Ontology. http://www.brenda-enzymes.org/ontology/tissue/tree/update/update_files/BrendaTissueOBO, 2011.

28. Briesemeister, S., T. Blum, S. Brady, Y. P. Lam, O. Kohlbacher, and H. Shatkay. 2009. SherLoc2: A high-accuracy hybrid method for predicting subcellular localization of proteins. *Journal of Proteome Research* 8 (11): 5363–5366.

29. Brill, E. 1995. Transformation-based error-driver learning and natural language processing: A case study in part-of-speech tagging. *Computational Linguistics* 21 (4): 543–565.

30. Bruford, E., M. Lush, M. Wright, T. Sneddon, S. Povey, and E. Birney. 2008. The HGNC database in 2008: A resource for the human genome. *Nucleic Acids Research* 36: D445–D448.

31. Bunescu, R., R. Ge, R. Kate, E. Marcotte, R. Mooney, A. Ramani, and Y. Wong. 2005. Comparative experiments on learning information extractors for proteins and their interactions. *Artificial Intelligence in Medicine* 33 (2): 139–155.

32. Cardie, C. 1997. Empirical methods in information extraction. *AI Magazine* 18 (4): 65–80.

33. Ceol, A., A. C. Aryamontri, L. Licata, D. Peluso, L. Briganti, L. Perfetto, L. Castagnoli, and G. Cesareni. 2010. MINT: A Molecular INTeraction database, 2009 update. *Nucleic Acids Research* (Database Issue) 38 (1): 532–539.

34. Chagoyen, M., P. Carmona-Saez, H. Shatkay, J. M. Caraz, and A. Pascual-Montano. 2006. Discovering semantic features in the literature: A foundation for building functional associations. *BMC Bioinformatics* 7: 41.

35. Chang, A., M. Scheer, A. Grote, I. Schomburg, and D. Schomburg. 2009. BRENDA, AMENDA and FRENDA the enzyme information system: New content and tools in 2009. *Nucleic Acids Research* 37: D588–D592.

36. Chang, J. T., S. Raychaudhuri, and R. B. Altman. Including biological literature improves homology search. In *Proceedings of the Pacific Symposium on Biocomputing*. Hackensack, NJ: World Scientific, 2001, 374–383.

37. Chen, D., H.-M. Muller, and P. W. Sternberg. 2006. Automatic document classification of biological literature. *BMC Bioinformatics* 7 (7): 370.

38. Chen, N., D. Blostein, and H. Shatkay. Exploring a new space of features for document classification: Figure clustering. In *Proceedings of the 2006 Conference of the Center for Advanced Studies on Collaborative Research*, 2006, 369–372.

39. Chen, N., D. Blostein, and H. Shatkay. Use of figures in literature mining for biomedical digital libraries. In *Proceedings of the Second IEEE International Conference on Document Image Analysis for Libraries*. Piscataway, NJ: IEEE, 2006, 180–197.

40. Chi, E. C., C. Friedman, N. Sager, and M. S. Lyman. Processing free-text input to obtain a database of medical information. In *Proceedings of the Eighth Annual International ACM SIGIR Conference on Research and Development in Information Retrieval*. New York: ACM, 1985, 82–90.

41. Cimino, J., G. Hripcsak, S. Johnson, and P. Clayton. Designing an introspective, multi-purpose, controlled medical vocabulary. In *Proceedings of the Annual Symposium on Computer Applications in Medical Care*, 1989, 513–518.

42. Coelho, L. P., A. Ahmed, A. Arnold, J. Kangas, A.-S. Sheikh, E. Xing, W. Cohen, and R. F. Murphy. Structured literature image finder: Extracting, information from text and images in biomedical literature. In Lecture Notes in Bioinformatics: Linking Literature, Information, and Knowledge for Biology, C. Blaschke, L. Hirschman, H. Shatkay, and A. Valencia, eds., vol. 6004, 23–32. New York: Springer, 2010.

43. Cohen, W. W., and Y. Singer. 1999. Context-sensitive learning methods for text categorization. *ACM Transactions on Information Systems* 17 (2): 141–173.

44. Collier, N., C. Nobata, and J. Tsujii. Extracting the names of genes and gene products with a hidden Markov model. In *Proceedings of the Eighteenth International Conference on Computational Linguistics*. Burlington, MA: Morgan Kaufmann, 2000, 201–207.

45. CoNLL-2010 Shared Task. http://www.inf.u-szeged.hu/rgai/conll2010st/, 2010.

46. Cowie, J., and W. Lehnert. 1996. Information extraction. *Communications of the ACM* 39 (1): 80–91.

47. Craven, M., and J. Kumlien. Constructing biological knowledge bases by extracting information from text sources. In *Proceedings of the Seventh International Conference on Intelligent Systems for Molecular Biology*. Menlo Park, CA: AAAI Press, 1999, 77–86.

48. Dai, H.-J., Y.-C. Chang, R. T.-H. Tsai, and W.-L. Hsu. 2011. Integration of gene normalization stages and co-reference resolution using a Markov-logic network. *Bioinformatics* 27 (18): 2586–2594.

49. Database of Interacting Proteins. http://dip.doe-mbi.ucla.edu/dip/Main.cgi, 2011.

50. de Bruijn, B., and J. Martin. 2002. Getting to the (c)ore of knowledge: Mining biomedical literature. *International Journal of Medical Informatics* 67: 7–18.

51. Deerwester, S., S. T. Dumais, G. Furnas, T. Landauer, and R. Harshman. 1990. Indexing by latent semantic analysis. *Journal of the Society for Information Science* 41 (6): 391–407.

52. DeGroot, M. H. 1986. *Probability and Statistics*. 2nd ed. Reading, MA: Addison-Wesley.

53. Demner-Fushman, D., S. Antani, and M. Simpson. 2010. *Combining Text and Visual Features for Biomedical Information Retrieval. A Report to the Board of Scientific Counselors*. Bethesda, MD: Lister Hill National Center for Biomedical Communications.

54. Denroche, R., R. Madupu, S. Yooseph, G. Sutton, and H. Shatkay. Toward computer-assisted text curation: Classification is easy (choosing training data can be hard . . .). In Lecture Notes in Bioinformatics: *Linking Literature, Information, and Knowledge for Biology*, C. Blaschke, L. Hirschman, H. Shatkay, and A. Valencia, eds., vol. 6004, 33–42. New York: Springer, 2010.

55. DiGiacomo, R., J. Kremer, and D. Shah. 1989. Fish-oil dietary supplementation in patients with Raynaud's phenomenon: A double-blind, controlled, prospective study. *American Journal of Medicine* 86 (2): 158–164.

56. Divoli, A., M. Wooldridge, and M. Hearst. 2010. Full text and figure display improves bioscience literature search. *PLoS ONE* 5 (4): e9619.

57. Dönnes, P., and A. Höglund. 2004. Predicting protein subcellular localization: Past, present, and future. *Genomics, Proteomics & Bioinformatics* 2 (4): 209–215.

58. Dowell, K., M. McAndrews-Hill, D. Hill, H. Drabkin, and J. Blake. Integrating text mining into the MGI biocuration workflow. *Database*, November 2009.

59. Draghici, S., P. Khatri, R. P. Martins, G. C. Ostermeier, and S. A. Krawetz. 2003. Global functional profiling of gene expression. *Genomics* 81 (2): 98–104.

60. Dumais, S. T. 1990. Enhancing performance in latent semantic indexing (LSI). *Behavior Research Methods, Instruments, & Computers* 23 (2): 229–236.

61. Dumais, S. T., G. W. Furnas, T. K. Landauer, S. Deerwester, and R. Harshman. Using latent semantic analysis to improve access to textual information. In *Proceedings of the ACM CHI 88 Human Factors in Computing Systems Conference*. New York: ACM, 1988, 281–285.

62. Dumais, S. T., J. Platt, D. Heckerman, and M. Sahami. Inductive learning algorithms and representations for text categorization. In *Proceedings of the 1998 ACM CIKM International Conference on Information and Knowledge Management*. New York: ACM, 1998, 148–155.

63. Elsevier Grand Challenge. http://www.elseviergrandchallenge.com/ index.html, 2009.

64. Farkas, R., V. Vincze, G. Móra, J. Csirik, and G. Szarvas. The CoNLL-2010 shared task: Learning to detect hedges and their scope in natural language text. In *Proceedings of the Fourteenth Conference on Computational Natural Language Learning: Shared Task*. Stroudsburg, PA: Association for Computational Linguistics, 2010, 1–12.

65. Folias, A., M. Matkovic, D. Bruun, S. Reid, J. Hejna, M. Grompe, A. D'Andrea, and R. Moses. 2002. BRCA1 interacts directly with the Fanconi anemia protein FANCA. *Human Molecular Genetics* 11 (21): 2591–2597.

66. Friedman, C., and G. Hripcsak. Evaluating natural language processors in the clinical domain. In *Proceedings of the Conference on Natural Language and Medical Concept Representation*, 1997, 41–52.

67. Friedman, C., P. Kra, H. Yu, M. Krauthammer, and A. Rzhetsky. 2001. GENIES: A natural-language processing system for the extraction of molecular pathways from journal articles. *Bioinformatics* 17 (suppl. 1): S74–S82.

68. Friedman, C., N. Sager, E. C. Chi, E. Marsh, C. Christenson, and M. S. Lyman. Computer structuring of free-text patient data. In *Proceedings of the Seventh Annual Symposium on Computer Applications in Medical Care,* 1983, 688–691.

69. Fukuda, K., T. Tsunoda, A. Tamura, and T. Takagi. Toward information extraction: Identifying protein names from biological papers. In *Pacific Symposium on Biocomputing*. Hackensack, NJ: World Scientific, 1998, 707–718.

70. Fundel, K., R. Küffner, and R. Zimmer. 2007. RelEx—relation extraction using dependency parse trees. *Bioinformatics* 23 (3): 365–371.

71. Furnas, G. W., S. Deerwester, S. T. Dumais, T. K. Landauer, R. A. Harshman, L. A. Streeter, and K. E. Lochbaum. Information retrieval using a singular value decomposition model of latent semantic structure. In *Proceedings of the Eleventh Annual International ACM SIGIR*

Conference on Research and Development in Information Retrieval. New York: ACM, 1988, 465–480.

72. Galperin, M. Y., and G. R. Cochrane. 2009. Nucleic Acids Research annual Database Issue and the NAR online Molecular Biology Database Collection in 2009. *Nucleic Acids Research* (Database Issue) 37: D1–D4.

73. Garten, Y., and R. B. Altman. 2009. Pharmspresso: A text mining tool for extraction of pharmacogenomic concepts and relationships from full text. Selected Proceedings of the First Summit on Translational Bioinformatics 2008. *BMC Bioinformatics* 10 (suppl. 2): S6.

74. Gene Ontology. http://www.geneontology.org/, 2011.

75. Gerner, M., G. Nenadic, and C. Bergman. 2010. LINNAEUS: A species name identification system for biomedical literature. *BMC Bioinformatics* 11: 85.

76. Glenisson, P., J. Mathys, and B. D. Moor. 2003. Meta-clustering of gene expression data and literature-extracted information. *ACM SIGKDD Explorations Special Issue on Microarray Data Mining* 5 (2): 101–112.

77. Gold, E. M. 1967. Language identification in the limit. *Information and Control* 10: 447–474.

78. Gold, E. M. 1978. Complexity of automaton identification from given data. *Information and Control* 37: 302–320.

79. Goldszmidt, M., and M. Sahami. A probabilistic approach to full-text document clustering. Technical Report ITAD-433-MS-98-044, SRI International, 1998.

80. Grishman, R., and B. Sundheim. Message Understanding Conference—6: A brief history. In *Proceedings of the Sixteenth International Conference on Computational Linguistics,* 1996, 466–471.

81. Grover, C., A. Lascarides, and M. Lapata. 2005. A comparison of parsing technologies for the biomedical domain. *Natural Language Engineering* 11 (1): 27–65.

82. Han, L., T. O. Suzek, Y. Wang, and S. H. Bryant. 2010. The text-mining based PubChem bioassay neighboring analysis. *BMC Bioinformatics* 11 (1): 549.

83. Hanisch, D., J. Fluck, H. Mevissen, and R. Zimmer. Playing biology's name game: Identifying protein names in scientific text. In *Proceedings of the Pacific Symposium on Biocomputing.* Hackensack, NJ: World Scientific, 2003, 403–414.

84. Hatzivassiloglou, V., P. Duboué, and A. Rzhetsky. 2001. Disambiguating proteins, genes, and RNA in text: A machine learning approach. *Bioinformatics* 17 (suppl. 1): S97–S106.

85. Hearst, M., A. Divoli, H. Guturu, A. Ksikes, P. Nakov, M. Wooldridge, and J. Ye. 2007. Biotext search engine: Beyond abstract search. *Bioinformatics* 23 (16): 2196–2197.

86. Hersh, W., and R. T. Bhupatiraju. TREC genomics track overview. In *Proceedings of the Text REtrieval Conference,* 2003.

87. Hersh, W., R. T. Bhupatiraju, L. Ross, P. Johnson, A. M. Cohen, and D. F. Kraemer. TREC 2004 genomics track overview. In *Proceedings of the Text REtrieval Conference,* 2004.

88. Hersh, W., C. Buckley, T. Leone, and D. Hickam. OHSUMED: An interactive retrieval evaluation and new large test collection for research. In *Proceedings of the Seventeenth Annual International ACM SIGIR Conference on Research and Development in Information Retrieval.* New York: ACM, 1994.

89. Hersh, W., A. M. Cohen, J. Yang, R. T. Bhupatiraju, P. Roberts, and M. Hearst. TREC 2005 genomics track overview. In *Proceedings of the Text REtrieval Conference,* 2005.

90. Hirschman, L., M. Colosimo, A. Morgan, and A. Yeh. 2005. Overview of BioCreative Task 1B: Normalized gene lists. *BMC Bioinformatics* 6 (suppl. 1): S11.

91. Hirschman, L., J. Park, J. Tsujii, L. Wong, and C. H. Wu. 2002. Accomplishments and challenges in literature data mining for biology. *Bioinformatics* 18: 1553–1561.

92. Hirschman, L., A. Yeh, C. Blaschke, and A. Valencia. 2005. Overview of BioCreative: Critical assessment of information extraction for biology. *BMC Bioinformatics* 6 (suppl. 1): S1.

93. Hoffmann, R., and A. Valencia. 2004. A gene network for navigating the literature. *Nature Genetics* 36 (7): 664.

94. Hoffmann, R., and A. Valencia. 2005. Implementing the iHOP concept for navigation of biomedical literature. *Bioinformatics* 21 (suppl. 2): ii252–ii258.

95. Hofmann, T. Probabilistic latent semantic indexing. In *Proceedings of the Twenty-Second Annual International ACM SIGIR Conference on Research and Development in Information Retrieval.* New York: ACM, 1999, 50–57.

96. Höglund, A., T. Blum, S. Brady, P. Dönnes, J. S. Miguel, M. Rocheford, O. Kohlbacher, and H. Shatkay. Significantly improved prediction of subcellular localization by integrating text and protein sequence data. In *Proceedings of the Pacific Symposium on Biocomputing.* Hackensack, NJ: World Scientific, 2006, 16–27.

97. Höglund, A., P. Dönnes, T. Blum, H. Adolph, and O. Kohlbacher. 2006. MultiLoc: Prediction of protein subcellular localization using n-terminal targeting sequences, sequence motifs, and amino acid composition. *BMC Bioinformatics* 22 (10): 1158–1165.

98. Hollingsworth, B., I. Lewin, and D. Tidhar. Retrieving hierarchical text structure from typeset scientific articles—a prerequisite for e-science text mining. In *Proceedings of the UK e-Science All Hands Meeting,* 2005.

99. Homayouni, R., K. Heinrich, L. Wei, and M. W. Berry. 2005. Gene clustering by latent semantic indexing of MEDLINE abstracts. *Bioinformatics* 21 (1): 104–115.

100. Hsu, C.-N., Y.-M. Chang, C.-J. Kuo, Y.-S. Lin, H.-S. Huang, and I.-F. Chung. 2008. Integrating high dimensional bi-directional parsing models for gene mention tagging. *Bioinformatics* 24 (13): i286–i294.

101. Hua, J., O. Ayasli, W. Cohen, and R. Murphy. Identifying fluorescence microscope images in online journal articles using both image and text features. In *Proceedings of the 2007 IEEE International Symposium on Biomedical Imaging.* Piscataway, NJ: IEEE, 2007, 1224–1227.

102. Hunter, L., Z. Lu, J. Firby, and W. Baumgartner, H. Johnson, P. Ogren, and K. B. Cohen. 2008. OpenDMAP: An open source, ontology-driven concept analysis engine with applications to capturing knowledge regarding protein transport, protein interactions and cell-type-specific gene expression. *BMC Bioinformatics* 9: 78.

103. iHOP. http://www.ihop-net.org/, 2010.

104. IntAct Molecular Interaction Database. http://www.ebi.ac.uk/intact/, 2010.

105. Iossifov, I., M. Krauthammer, C. Friedman, V. Hatzivassilogou, J. Bader, K. White, and A. Rzhetsky. 2004. GeneWays: A system for extracting, analyzing, visualizing, and integrating molecular pathway data. *Journal of Biomedical Informatics* 37 (1): 43–53.

106. Jain, N., C. Knirsch, C. Friedman, and G. Hripcsak. Identification of suspected tuberculosis patients based on natural language processing of chest radiograph reports. In *Proceedings of the Annual Symposium of the American Medical Informatics Association,* 1996, 542–546.

107. JCVI CHAR Database. http://www.jcvi.org/cms/research/projects/algorithmically -tuned-protein-families-rule-base-and-characterized-proteins/overview/, 2011.

108. JCVI. Comprehensive Microbial Resource. http://cmr.jcvi.org/tigr-scripts/CMR/ CmrHomePage.cgi, 2011.

109. Jenssen, T.-K., A. Laegreid, J. Komorowski, and E. Hovig. 2001. A literature network of human genes for high-throughput analysis of gene expression. *Nature Genetics* 28: 21–28.

110. Jiang, F., and M. Littman. Approximate dimension equalization in vector-based information retrieval. In *Proceedings of Seventeenth International Conference on Machine Learning.* Burlington, MA: Morgan Kaufmann, 2000, 423–430.

111. Jiang, J., and C. Zhai. 2007. An empirical study of tokenization strategies for biomedical information retrieval. *Information Retrieval* 10 (4–5): 341–363.

112. Joachims, T. A probabilistic analysis of the Rocchio algorithm with TFIDF for text categorization. In *Proceedings of the Fourteenth International Conference on Machine Learning*. Burlington, MA: Morgan Kaufmann, 1997, 143–151.

113. Joachims, T. Text categorization with support vector machines: Learning with many relevant features. In *Proceedings of the Tenth European Conference on Machine Learning*. New York: Springer, 1998, 137–142.

114. Johnson, H., K. B. Cohen, W. A. Baumgartner, Z. Lu, M. Bada, T. Kester, H. Kim, and L. Hunter. Evaluation of lexical methods for detecting relationships between concepts from multiple ontologies. In *Proceedings of the Pacific Symposium on Biocomputing*. Hackensack, NJ: World Scientific, 2006, 28–39.

115. Jurafsky, D., and J. Martin. 2009. *Speech and Language Processing: An Introduction to Natural Language Processing, Speech Recognition, and Computational Linguistics*. 2nd ed. Upper Saddle River, NJ: Prentice Hall.

116. Kang, N., E. M. van Mulligan, and J. A. Kors. 2011. Comparing and combining chunkers of biomedical text. *Journal of Biomedical Informatics* 44 (2): 354–360.

117. Karamanis, N., I. Lewin, R. Seal, R. Drysdale, and E. J. Briscoe. Integrating natural language processing with FlyBase curation. In *Proceedings of the Pacific Symposium on Biocomputing*. Hackensack, NJ: World Scientific, 2007, 245–256.

118. Karamanis, N., R. Seal, I. Lewin, P. McQuilton, A. Vlachos, C. Gasperin, R. Drysdale, and T. Briscoe. 2008. Natural language processing in aid of FlyBase curators. *BMC Bioinformatics* 9: 193.

119. KDD Cup 2002. http://www.biostat.wisc.edu/~craven/kddcup/.

120. Khatri, P., and S. Draghici. 2005. Ontological analysis of gene expression data: Current tools, limitations, and open problems. *Bioinformatics* 21: 3587–3595.

121. Kim, J.-D., T. Ohta, S. Pyysalo, Y. Kano, and J. Tsujii. Overview of BioNLP'09 shared task on event extraction. In *Proceedings of the Workshop on BioNLP: Shared Task*. Stroudsburg, PA: Association for Computational Linguistics, 2009.

122. Kim, J.-D., S. Pyysalo, T. Ohta, R. Bossy, N. Nguyen, and J. Tsujii. Overview of BioNLP shared task 2011. In *Proceedings of the BioNLP Shared Task 2011 Workshop*. Stroudsburg, PA: Association for Computational Linguistics, 2011.

123. Klein, T., J. Chang, M. Cho, K. Easton, R. Fergerson, M. Hewett, Z. Lin, et al. 2001. Integrating genotype and phenotype information: An overview of the PharmGKB project. *Pharmacogenomics Journal* 1 (3): 167–170.

124. Klinger, R., C. Kolarik, J. Fluck, M. Hofmann-Apitius, and C. Friedrich. 2008. Detection of IUPAC and IUPAC-like chemical names. *Bioinformatics* 24 (13): i268–i276.

125. Kou, Z., W. Cohen, and R. Murphy. A stacked graphical model for associating information from text and images in figures. In *Proceedings of the Pacific Symposium on Biocomputing*. Hackensack, NJ: World Scientific, 2007, 257–268.

126. Krallinger, M., F. Leitner, C. Rodriguez-Penagos, and A. Valencia. 2008. Overview of the protein-protein interaction annotation extraction task of BioCreative II. *Genome Biology* 9 (suppl. 2): S4.

127. Krallinger, M., A. Morgan, L. Smith, F. Leitner, L. Tanabe, J. Wilbur, L. Hirschman, and A. Valencia. 2008. Evaluation of text-mining systems for biology: overview of the Second BioCreativecommunity challenge. *Genome Biology* 9 (suppl. 2): S1.

128. Krallinger, M., and A. Valencia. 2005. Applications of text mining in molecular biology, from name recognition to protein interaction maps. In *Data Analysis and Visualization in Genomics and Proteomics*, F. Azuaje and J. Dopazo, eds., Hoboken, NJ: Wiley.

129. Krauthammer, M., P. Kra, I. Iossifov, S. Gomez, G. Hripcsak, V. Hatzivassiloglou, C. Friedman, and A. Rzhetsky. 2002. Of truth and pathways: Chasing bits of information through myriads of articles. *Bioinformatics* 18 (suppl. 1): S249–S257.

130. Küffner, R., K. Fundel, and R. Zimmer. 2005. Expert knowledge without the expert: Integrated analysis of gene expression and literature to derive active functional contexts. *Bioinformatics* 21: 259–267.

131. Lafferty, J., A. McCallum, and F. Pereira. Conditional random fields: Probabilistic models for segmenting and labeling sequence data. In *Proceedings of the Eighteenth International Conference on Machine Learning.* Burlington, MA: Morgan Kaufmann, 2001, 282–289.

132. Larkey, L. S., and W. B. Croft. Combining classifiers in text categorization. In *Proceedings of Nineteenth Annual International ACM SIGIR Conference on Research and Development in Information Retrieval.* New York: ACM, 1996, 289–297.

133. Lee, W.-J., L. Raschid, P. Srinivasan, N. Shah, D. Rubin, and N. Noy. Using annotations from controlled vocabularies to find meaningful associations. In *Proceedings of the Workshop on Data Integration in the Life Sciences,* 2007.

134. Leek, T. Information extraction using hidden Markov models. Master's thesis, Department of Computer Science and Engineering, University of California, San Diego, 1997.

135. Leitner, F., S. A. Mardis, M. Krallinger, G. Cesareni, L. A. Hirschman, and A. Valencia. 2010. An overview of BioCreative II.5. *IEEE/ACM Transactions on Computational Biology and Bioinformatics* 7 (3): 385–399.

136. Letunic, I., T. Doerks, and P. Bork. 2009. SMART 6: Recent updates and new developments. *Nucleic Acids Research* 37: D229–D232.

137. Lewis, D. D. Evaluating and optimizing autonomous text classification systems. In *Proceedings of the Eighteenth Annual International ACM SIGIR Conference on Research and Development in Information Retrieval, SIGIR-95.* New York: ACM, 1995, 246–254.

138. Lewis, D. D., and P. J. Hayes. 1994. Guest editorial for the special issue on text categorization. *ACM Transactions on Information Systems* 12 (3): 231.

139. Lewis, D. D., and M. Ringuette. A comparison of two learning algorithms for text categorization. In *Proceedings of the Third Annual Symposium on Document Analysis and Information Retrieval,* 1994, 81–93.

140. Lewis, D. D., R. E. Schapire, J. P. Callan, and R. Papka. Training algorithms for linear text classifiers. In *Proceedings of the Nineteenth Annual International ACM SIGIR Conference on Research and Development in Information Retrieval,* New York: ACM, 1996, 298–306.

141. Lewis, D. D., Y. Yang, T. Rose, and F. Li. 2004. RCV1: A new benchmark collection for text categorization research. *Journal of Machine Learning Research* 5: 361–397.

142. Li, Y., H. Lin, and Z. Yang. 2009. Incorporating rich background knowledge for gene named entity classification and recognition. *BMC Bioinformatics* 10: 223.

143. Light, M., X. Quiu, and P. Srinivasan. The language of bioscience: Facts, speculations, and statements in between. In *Proceedings of the BioLINK Workshop on Linking Biological Literature, Ontologies and Databases,* 2004.

144. Lindberg, D. A., B. L. Humphreys, and A. T. McCray. 1993. The Unified Medical Language System. *Methods of Information in Medicine* 32 (4): 281–291.

145. Lowe, H. J., and G. O. Barnett. 1994. Understanding and using the Medical Subject Headings (MeSH) vocabulary to perform literature searches. *Journal of the American Medical Association* 271 (14): 1103–1108.

146. Lu, Z., H.-Y. Kao, C.-H. Wei, M. Huang, J. Liu, C.-J. Kuo, C.-N. Hsu, R. T.-H. Tsai, H.-J. Dai, N. Okazaki, H.-C. Cho, M. Gerner, I. Solt, S. Agarwal, F. Liu, D. Vishnyakova, P. Ruch, M. Romacker, F. Rinaldi, S. Bhattacharya, P. Srinivasan, H. Liu, M. Torii, S. Matos, D. Campos, K. Verspoor, K. M. Livingston, and W. J. Wilbur. 2011. The gene normalization task in BioCreative III. *BMC Bioinformatics* 12 (suppl. 8): S2.

147. Manning, C., and P. Raghavan. 2008. *Introduction to Information Retrieval.* Cambridge, UK: Cambridge University Press.

148. Manning, C., and H. Schütze. 1999. *Foundations of Statistical Natural Language Processing.* Cambridge, MA: MIT Press.

149. Marcotte, E., I. Xenarios, and D. Eisenberg. 2000. Mining literature for protein-protein interactions. *Bioinformatics* 17: 359–363.

150. Marcus, M. P., B. Santorini, and M. A. Marcinkiewicz. 1993. Building a large annotated corpus of English: The Penn Treebank. *Computational Linguistics* 19 (2): 313–330.

151. McCallum, A., and K. Nigam. A comparison of event models for naive Bayes text classification. In *Proceedings of the AAAI/ICML Workshop on Learning for Text Categorization,* 1998, 41–48.

152. McCallum, A. K., and K. Nigam. Text classification by bootstrapping with keywords, EM and shrinkage. In *Proceedings of the Workshop on Unsupervised Learning in Natural Language Processing,* 1999.

153. McClosky, D., E. Charniak, and M. Johnson. Automatic domain adaptation for parsing. In *Proceedings of Human Language Technologies: The 2010 Annual Conference of the North American Chapter of the Association for Computational Linguistics.* Stroudsburg, PA: Association for Computational Linguistics, 2010, 28–36.

154. McClosky, D., M. Surdeanu, and C. Manning. Event extraction as dependency parsing. In *Proceedings of the Forty-Ninth Annual Meeting of the Association for Computational Linguistics: Human Language Technologies.* Stroudsburg, PA: Association for Computational Linguistics, 2011.

155. McNamee, P., C. Nicholas, and J. Mayfield. Addressing morphological variation in alphabetic languages. In *Proceedings of the Thirty-Second Annual International ACM SIGIR Conference on Research and Development in Information Retrieval.* New York: ACM, 2009, 75–82.

156. Medlock, B. 2008. Exploring hedge identification in biomedical literature. *Journal of Biomedical Informatics* 41 (4): 636–654.

157. MeSH: Medical Subject Headings. http://www.nlm.nih.gov/mesh/, 2011.

158. MGI. Mouse Genome Informatics. 2011. http://www.informatics.jax.org/

159. MINT. The Molecular Interaction Database. http://mint.bio.uniroma2.it/mint/, 2011.

160. Mitchell, T. 1997. *Machine Learning.* New York: McGraw-Hill.

161. Mitchell, T. *Machine Learning.* Web revisions towards 2nd edition, http://www.cs.cmu.edu/People/tom/mlbook/NBayesLogReg.pdf, 2005.

162. Miwa, M., R. Saetre, Y. Miyao, and J. Tsujii. 2009. Protein-protein interaction extraction by leveraging multiple kernels and parsers. *International Journal of Medical Informatics* 78 (12): e39–e46.

163. Morgan, A., L. Hirschman, M. Colosimo, A. Yeh, and J. Colombe. 2004. Gene name identification and normalization using a model organism database. *Journal of Biomedical Informatics* 37 (6): 396–410.

164. Morgan, A., Z. Lu, X. Wang, A. Cohen, J. Fluck, P. Ruch, A. Divoli, et al. 2008. Overview of BioCreative II gene normalization. *Genome Biology* 9 (S2): S3.

165. Muller, H.-M., E. E. Kenny, and P. W. Sternberg. 2004. Textpresso: An ontology-based information retrieval and extraction system for biological literature. *PLoS Biology* 2 (11): e309.

166. Muller, H.-M., A. Rangarajan, T. K. Teal, and P. W. Sternberg. 2008. Textpresso for neuroscience: Searching the full text of thousands of neuroscience research papers. *Neuroinformatics* 6 (3): 195–204.

167. Murphy, R. F., Z. Kou, J. Hua, M. Joffe, and W. W. Cohen. Extracting and structuring subcellular location information from on-line journal articles: The subcellular location image finder. In *Proceedings of the IASTED International Conference on Knowledge Sharing and Collaborative Engineering.* Calgary, Alberta, Canada: ACTA Press, 2004.

168. Murphy, R. F., M. Velliste, J. Yao, and G. Porreca. Searching online journals for fluorescence microscope images depicting protein subcellular location patterns. In *Proceedings of the IEEE International Symposium on Bioinformatics and Biomedical Engineering.* Piscataway, NJ: IEEE, 2001, 119–128.

169. Nair, R., and B. Rost. 2002. Inferring sub-cellular localization through automated lexical analysis. *Bioinformatics* 18 (suppl. 1): S78–S86.

170. Naoaki, O., H.-C. Cho, R. Sætre, S. Pyysalo, T. Ohta, and J. Tsujii. The gene normalization and interactive systems of the University of Tokyo in the BioCreative III challenge. In *Proceedings of the BioCreative III Workshop,* 2010, 125–130.

171. Entrez, http://www.ncbi.nlm.nih.gov/Entrez/, 2011.

172. NCBI. Taxonomy Database. http://www.ncbi.nlm.nih.gov/Taxonomy/, 2011.

173. Newton, M. A., F. A. Quintana, J. A. Den, S. Sengupta, P. Ahlquist, and C. Chile. 2007. Random-set methods identify distinct aspects of the enrichment signal in gene-set analysis. *Annals of Applied Statistics* 1 (1): 85–106.

174. Ng, A., and M. Jordan. 2002. On discriminative vs. generative classifiers: A comparison of logistic regression and naive Bayes. In *Advances in Neural Information Processing Systems.* vol. 14. G. Tesauro, D. Touretzky, and T. Leen, eds., Cambridge, MA: MIT Press.

175. Nguyen, N., J.-D. Kim, and J. Tsujii. Challenges in pronoun resolution system for biomedical text. In *Proceedings of the Sixth International Conference on Language Resources and Evaluation.* Paris: European Language Resources Association, 2008, 26–28.

176. Ogren, P. V., K. B. Cohen, G. K. Acquaah-Mensah, J. Eberlein, and L. Hunter. The compositional structure of Gene Ontology terms. In *Proceedings of the Pacific Symposium on Biocomputing.* Hackensack, NJ: World Scientific, 2004, 214–225.

177. Pafilis, E., S. O'Donoghue, L. J. Jensen, H. Horn, M. Kuhn, N. P. Brown, and R. Schneider. 2009. Reflect: Augmented browsing for the life scientist. *Nature Biotechnology* 27 (6): 508–510.

178. Papadimitriou, C. H., P. Raghavan, H. Tamaki, and S. Vempala. 2000. Latent semantic indexing: A probabilistic analysis. *Journal of Computer and System Sciences* 61 (2): 217–235.

179. Peterson, J. D., L. A. Umayam, T. M. Dickinson, E. K. Hickey, and O. White. 2001. The comprehensive microbial resource. *Nucleic Acids Research* 29 (1): 123–125.

180. Ponte, J. M., and W. B. Croft. A language modeling approach to information retrieval. In *Proceedings of the Twenty-First Annual International ACM SIGIR Conference on Research and Development in Information Retrieval,* New York: ACM, 1998, 275–281.

181. Poon, H., and L. Vanderwende. Joint inference for knowledge extraction from the biomedical literature. In *Proceedings of Human Language Technologies: The 2010 Annual Conference of the North American Chapter of the Association for Computational Linguistics.* Stroudsburg, PA: Association for Computational Linguistics, 2010, 813–821.

182. Porter, M. F. 1980. An algorithm for suffix stripping. *Program* 14 (3): 127–130.

183. Protein Data Bank. http://www.rcsb.org/pdb/, 2011.

184. PubChem. http://pubchem.ncbi.nlm.nih.gov/, 2011.

185. PubGene. http://www.pubgene.org/, 2011.

186. PubMed. http://www.ncbi.nlm.nih.gov/pubmed/, 2011.

187. PubMed Central. http://www.ncbi.nlm.nih.gov/pmc/, 2011.

188. Qian, Y., and R. Murphy. 2008. Improved recognition of figures containing fluorescence microscope images in online journal articles using graphical models. *Bioinformatics* 24: 569–576.

189. Quinlan, J. 1993. *C4.5: Programs for Machine Learning.* Burlington, MA: Morgan Kaufmann.

190. Rabiner, L. R. 1989. A tutorial on hidden Markov models and selected applications in speech recognition. *Proceedings of the IEEE* 77 (2): 257–286.

191. Rafkind, B., M. Lee, S. Chang, and H. Yu. Exploring text and image features to classify images in bioscience literature. In *Proceedings of the BioNLP Workshop on Linking Natural Language Processing and Biology,* 2006.

192. Ramani, A., R. Bunescu, R. Mooney, and E. Marcotte. 2005. Consolidating the set of known human protein-protein interactions in preparation for large-scale mapping of the human interactome. *Genome Biology* 6: R40.

193. Ratnaparkhi, A. A maximum entropy model for part-of-speech tagging. In *Proceedings of the Empirical Methods in Natural Language Processing Conference.* Stroudsburg, PA: Association for Computational Linguistics, 1996, 133–142.

194. Ray, S., and M. Craven. Representing sentence structure in hidden Markov models for information extraction. In *Proceedings of the Seventeenth International Joint Conference on Artificial Intelligence.* Burlington, MA: Morgan Kaufmann, 2001, 1273–1279.

195. Reflect. http://reflect.ws/, 2011.

196. Regev, Y., M. Finkelstein-Landau, R. Feldman, M. Gorodetsky, X. Zheng, S. Levy, R. Charlab, et al. 2002. Rule-based extraction of experimental evidence in the biomedical domain—the KDD Cup (task 1). *SIGKDD Explorations* 4 (2): 90–91.

197. Regev, Y., M. Finkelstein-Landau, R. Feldman, X. Zheng, S. Levy, R. Charlab, C. Lawrence, R. A. Lippert, Q. Zhang, and H. Shatkay. December 2002. Rule-based extraction of experimental evidence in the biomedical domain—the KDD Cup 2002 (task 1). *SIGKDD Explorations* 4 (2): 90–92.

198. Renner, A., and A. Aszódi. High-throughput functional annotation of novel gene products using document clustering. In *Proceedings of the Pacific Symposium on Biocomputing.* Hackensack, NJ: World Scientific, 2000, 54–65.

199. Reuters-21578 Test Collections. http://www.daviddlewis.com/resources/testcollections/reuters21578, 2011.

200. Riedel, S., and A. McCallum. Fast and robust joint models for biomedical event extraction. In *Proceedings of the 2011 Conference on Empirical Methods in Natural Language Processing.* Stroudsburg, PA: Association for Computational Linguistics, 2011, 1–12.

201. Riloff, E., and W. Lehnert. 1994. Information extraction as a basis for high-precision text classification. *ACM Transactions on Information Systems* 12: 296–333.

202. Rinaldi, F., S. Clematide, G. Schneider, M. Romacker, and T. Vachon. ODIN: An advanced interface for the curation of biomedical literature. *Nature Precedings,* http://precedings.nature.com/documents/5169/version/1, 2010.

203. Roberts, P. M., A. M. Cohen, and W. R. Hersh. 2009. Tasks, topics and relevance judging for the TREC Genomics track: Five years of experience evaluating biomedical text information retrieval systems. *Information Retrieval* 12 (1): 81–97.

204. Russell, S., and P. Norvig. 2002. *Artificial Intelligence: A Modern Approach.* Upper Saddle River, NJ: Prentice Hall.

205. Saccharomyces Genome Database. http://www.yeastgenome.org/, 2011.

206. Salton, G. 1989. *Automatic Text Processing.* Reading, MA: Addison-Wesley.

207. Sayers, E., T. Barrett, D. Benson, S. Bryant, K. Canese, V. Chetvernin, D. Church, et al. 2009. Database resources of the National Center for Biotechnology Information. *Nucleic Acids Research* 37: D5–D15.

208. Schütze, H., D. A. Hull, and J. O. Pedersen. A comparison of classifiers and document representations for the routing problem. In *Proceedings of the Eighteenth Annual International ACM SIGIR Conference on Research and Development in Information Retrieval.* New York: ACM, 1995, 229–237.

209. Sebastiani, F. 2002. Machine learning in automated text categorization. *ACM Computing Surveys* 34 (1): 1–47.

210. Settles, B. 2005. ABNER: An open source tool for automatically tagging genes, proteins, and other entity names in text. *Bioinformatics* 21 (14): 3191–3192.

211. Settles, B. Active learning literature survey. Computer Sciences Technical Report 1648, Department of Computer Sciences, University of Wisconsin-Madison, 2010.

212. Shatkay, H., N. Chen, and D. Blostein. 2006. Integrating image data into biomedical text categorization. *Bioinformatics* 22 (14): e446–e453.

213. Shatkay, H., S. Edwards, W. J. Wilbur, and M. Boguski. Genes, themes and microarrays: Using information retrieval for large scale gene analysis. In *Proceedings of the Eighth International Conference on Intelligent Systems for Molecular Biology*. Menlo Park, CA: AAAI Press, 2000, 317–328.

214. Shatkay, H., and R. Feldman. 2003. Mining the biomedical literature in the genomic era: An overview. *Journal of Computational Biology* 10 (6): 821–855.

215. Shatkay, H., A. Höglund, S. Brady, T. Blum, P. Dönnes, and O. Kohlbacher. 2007. SherLoc: High-accuracy prediction of protein subcellular localization by integrating text and protein sequence data. *Bioinformatics* 23 (11): 1410–1417.

216. Shatkay, H., F. Pan, A. Rzhetsky, and W. J. Wilbur. 2008. Multi-dimensional classification of biomedical text: Toward automated, practical provision of high-utility text to diverse users. *Bioinformatics* 24 (18): 2086–2093.

217. Shatkay, H., and W. J. Wilbur. Finding themes in MEDLINE documents: Probabilistic similarity search. In *Proceedings of the IEEE Conference on Advances in Digital Libraries*. Piscataway, NJ: IEEE, 2000, 183–192.

218. Shaw, W. M., R. Burgin, and P. Howell. 1997. Performance standards and evaluations in IR test collections: Cluster-based retrieval models. *Information Processing & Management* 33 (1): 1–14.

219. Shaw, W. M., R. Burgin, and P. Howell. 1997. Performance standards and evaluations in IR test collections: Vector-space and other retrieval models. *Information Processing & Management* 33 (1): 15–36.

220. Shen, D., J. Zhang, G. Zhou, J. Su, and C.-L. Tan. Effective adaptation of hidden Markov model-based named entity recognizer for biomedical domain. In *Proceedings of the ACL Workshop on Natural Language Processing in Biomedicine*. Stroudsburg, PA: Association for Computational Linguistics, 2003.

221. Skounakis, M., M. Craven, and S. Ray. Hierarchical hidden Markov models for information extraction. In *Proceedings of the Eighteenth International Joint Conference on Artificial Intelligence*. Burlington, MA: Morgan Kaufmann, 2003, 427–433.

222. Smalheiser, N. R., and V. I. Torvik. 2008. The place of literature-based discovery in contemporary scientific practice. In *Literature-Based Discovery*. vol. 15. M. Weeber and P. Bruza, eds., New York: Springer, 13–22.

223. Smith, B., M. Ashburner, C. Rosse, J. Bard, W. Bug, W. Ceusters, L. Goldberg, K. Eilbeck, A. Ireland, and C. Mungall, N. Leontis, P. Rocca-Serra, A. Ruttenberg, S.-A. Sansone, R. Scheurmann, N. Shah, P. Whetzel, and S. Lewis. 2007. The OBO foundry: Coordinated evolution of ontologies to support biomedical data integration. *Nature Biotechnology* 25 (11): 1251–1255.

224. Smith, L., T. Rindflesch, and W. J. Wilbur. 2004. MedPost: A part-of-speech tagger for biomedical text. *Bioinformatics* 20 (14): 2320–2321.

225. Smith, L., L. Tanabe, R. Ando, C. Kuo, I. Chung, C. Hsu, Y. Lin, et al. 2008. Overview of BioCreative II gene mention recognition. *Genome Biology* 9 (suppl. 2): S2.

226. Spärck-Jones, K., S. Walker, and S. Robertson. 2000. A probabilistic model of information retrieval: Development and status. *Information Processing & Management* 36 (6): 779–840.

227. Srinivasan, P. 2004. Text mining: Generating hypotheses from MEDLINE. *Journal of the American Society for Information Science American Society for Information Science* 55 (5): 396–413.

228. Srinivasan, P., and B. Libbus. 2004. Mining MEDLINE for implicit links between dietary substances and diseases. *Bioinformatics* 20 (suppl. 1): i290–i296.

229. Stapley, B. J., and G. Benoit. Bibliometrics: Information retrieval and visualization from co-occurrences of gene names in MEDLINE abstracts. In *Proceedings of the Pacific Symposium on Biocomputing.* Hackensack, NJ: World Scientific, 2000, 526–537.

230. Stapley, B. J., L. A. Kelley, and M. J. E. Sternberg. Predicting the subcellular location of proteins from text using support vector machines. In *Proceedings of the Pacific Symposium on Biocomputing.* Hackensack, NJ: World Scientific, 2002, 374–385.

231. Stark, C., B. Breitkreutz, T. Reguly, L. Boucher, A. Breitkreutz, and M. Tyers. 2006. BioGRID: A general repository for interaction datasets. *Nucleic Acids Research* 34 (Database Issue): 535–539.

232. STITCH. http://stitch.embl.de/, 2011.

233. Sutton, C., and A. McCallum. 2006. An introduction to conditional random fields for relational learning. In *Introduction to Statistical Relational Learning*, L. Getoor and B. Taskar, eds., Cambridge, MA: MIT Press.

234. Swanson, D. R. 1986. Fish-oil, Raynaud's syndrome and undiscovered public knowledge. *Perspectives in Biology and Medicine* 30 (1): 7–18.

235. Swanson, D. R. 1986. Undiscovered public knowledge. *Library Quarterly* 56 (2): 103–118.

236. Swanson, D. R. 1988. Migraine and magnesium: Eleven neglected connections. *Perspectives in Biology and Medicine* 31 (4): 526–557.

237. Swanson, D. R. 1990. Somatomedin C and Arginine: Implicit connections between mutually isolated literatures. *Perspectives in Biology and Medicine* 33 (2): 157–186.

238. Swanson, D. R., and N. R. Smalheiser. 1997. An interactive system for finding complementary literatures: A stimulus to scientific discovery. *Artificial Intelligence* 91: 183–203.

239. Swanson, D. R., and N. R. Smalheiser. 1999. Implicit text linkage between MEDLINE records: Using Arrowsmith as an aid to scientific discovery. *Library Trends* 48 (1): 48–59.

240. Swanson, D. R., N. R. Smalheiser, and A. Bookstein. 2001. Information discovery from complementary literatures: Categorizing viruses as potential weapons. *Journal of the American Society for Information Science and Technology* 52 (10): 797–812.

241. TAIR. The Arabidopsis Information Resource. http://www.arabidopsis.org/, 2011.

242. Tateisi, Y., A. Yakushiji, T. Ohta, and J. Tsujii. 2005. Syntax annotation for the GENIA corpus. In *Natural Language Processing—IJCNLP 2005*, R. Dale, K.-F. Wong, J. Su, and O. Y. Kwong, eds., New York: Springer, 222–227.

243. Text REtrieval Conference (TREC). http://trec.nist.gov, 2011.

244. Textpresso. http://www.textpresso.org/, 2011.

245. The FlyBase Consortium. 2002. The FlyBase database of the *Drosophila* genome projects and community literature. *Nucleic Acids Research* 30 (1): 106–108.

246. The Gene Ontology Consortium. 2000. Gene Ontology: Tool for the unification of biology. *Nature Genetics* 25: 25–29.

247. The Gene Ontology Consortium. 2010. The Gene Ontology in 2010: Extensions and refinements. *Nucleic Acids Research* 38 (Database Issue): D331–D335.

248. The UniProt Consortium. 2009. The universal protein resource (UniProt). *Nucleic Acids Research* 37(Database Issue): D169–D174.

249. Thoma, G. 2001. *Automating the Production of Bibliographic Records for MEDLINE. Technical Report.* Bethesda, MD: National Library of Medicine.

250. Thomas, P. E., J. Starlinger, C. Jacob, I. Solt, J. Hakenberg, and U. Leser. Geneview gene-centric ranking of biomedical text. In *Proceedings of the BioCreative III Workshop,* 2010, 137–142.

251. LingPipe. http://alias-i.com/lingpipe/.

252. Toutanova, K., D. Klein, C. Manning, and Y. Singer. Feature-rich part-of-speech tagging with a cyclic dependency network. In *Proceedings of the 2003 Human Language Technology Conference of the North American Chapter of the Association for Computational Linguistics.* Stroudsburg, PA: Association for Computational Linguistics, 2003, 252–259.

253. TREC. Genomics Track. http://ir.ohsu.edu/genomics, 2008.

254. Tsuruoka, Y., Y. Tateishi, J.-D. Kim, T. Ohta, J. McNaught, S. Ananiadou, and J. Tsujii. 2005. Developing a robust part-of-speech tagger for biomedical text. In *Advances in Informatics.* vol. 3746. P. Bozanis and E. Houstis, eds., New York: Springer, 382–392.

255. Twigger, S., M. Shimoyama, S. Bromberg, A. Kwitek, and H. Jacob, and the RGD Team. 2007. The Rat Genome Database, update 2007—easing the path from disease to data and back again. *Nucleic Acids Research* 35 (Database Issue): D658–D662.

256. Unified Medical Language System. http://www.nlm.nih.gov/research/umls/, 2011.

257. UniProt. http://www.uniprot.org/, 2011.

258. Van Auken, K., J. Jaffery, J. Chan, H.-M. Muller, and P. W. Sternberg. 2009. Semi-automated curation of protein subcellular localization: A text mining-based approach to gene ontology (GO) cellular component curation. *BMC Bioinformatics* 10: 228.

259. van Rijsbergen, C. J. 1977. A theoretical basis for the use of co-occurrence data in information retrieval. *Journal of Documentation* 33 (2): 106–119.

260. van Rijsbergen, C. J. 1979. *Information Retrieval.* London: Butterworth.

261. Vapnik, V. 1995. *The Nature of Statistical Learning Theory.* New York: Springer.

262. Vlachos, A., and M. Craven. Detecting speculative language using syntactic dependencies and logistic regression. In *Proceedings of the Fourteenth Conference on Computational Natural Language Learning.* Stroudsburg, PA: Association for Computational Linguistics, 2010, 18–25.

263. Vlachos, A., and M. Craven. Search-based structured prediction applied to biomedical event extraction. In *Proceedings of the Fifteenth Conference on Computational Natural Language Learning.* Stroudsburg, PA: Association for Computational Linguistics, 2011, 49–57.

264. Weeber, M., H. Klein, L. T. de Jong-van den Berg, and R. Vos. 2001. Using concepts in literature-based discovery: Simulating Swanson's Raynaud-fish oil and migraine-magnesium discoveries. *Journal of the American Society for Information Science* 52 (7): 548–557.

265. Wilbur, W. J. 1992. An information measure of retrieval performance. *Information Systems* 17 (4): 283–298.

266. Wilbur, W. J., and Y. Yang. 1996. An analysis of statistical term strength and its use in the indexing and retrieval of molecular biology text. *Computers in Biology and Medicine* 26 (3): 209–222.

267. Witten, I. H., A. Moffat, and T. C. Bell. 1999. *Managing Gigabytes, Compressing and Indexing Documents and Images.* 2nd ed. Burlington, MA: Morgan Kaufmann.

268. WormBase. http://www.wormbase.org/, 2011.

269. Wren, J. D., R. Bekeredjian, J. A. Stewart, R. V. Shohet, and H. R. Garner. 2004. Knowledge discovery by automated identification and ranking of implicit relationships. *Bioinformatics* 20 (3): 389398.

270. Xenarios, I., D. Rice, L. Salwinski, M. Baron, E. Marcotte, and D. Eisenberg. 2000. DIP: The database of interacting proteins. *Nucleic Acids Research* 28: 289–291.

271. Xu, S., J. McCusker, and M. Krauthammer. Exploring the use of image text for biomedical literature retrieval. In *Proceedings of the Annual Symposium of the American Medical Informatics Association,* 2008.

272. Xu, S., J. McCusker, and M. Krauthammer. 2008. Yale image finder (YIF): A new search engine for retrieving biomedical images. *Bioinformatics* 24 (17): 1968–1970.

273. Xuan, W., S. Watson, and F. Meng. 2007. Tagging sentence boundaries in biomedical literature. In *Computational Linguistics and Intelligent Text Processing*. vol. 4394. A. Gelbukh, ed., New York: Springer, 186–195.

274. Yang, Y. 1999. An evaluation of statistical approaches to text categorization. *Information Retrieval* 1 (1-2): 69–90.

275. Yang, Y. A study of thresholding strategies for text categorization. In *Proceedings of the Twenty-Fourth Annual International ACM SIGIR Conference on Research and Development in Information Retrieval*. New York: ACM, 2001, 137–145.

276. Yang, Y., and C. G. Chute. 1994. An example-based mapping method for text categorization and retrieval. *ACM Transactions on Information Systems* 12 (3): 252–277.

277. Yang, Y., and X. Liu. A re-examination of text categorization methods. In *Proceedings of the Twenty-Second Annual International ACM SIGIR Conference on Research and Development in Information Retrieval*. New York: ACM, 1999, 42–49.

278. Yeh, A., L. Hirschman, and A. Morgan. 2002. Background and overview for KDD Cup 2002 task 1: Information extraction from biomedical articles. *SIGKDD Explorations* 4 (2): 87–89.

279. Yeh, A., A. Morgan, M. Colosimo, and L. Hirschman. (May 2005). BioCreative Task 1a: Gene mention finding evaluation. *BMC Bioinformatics* 6 (suppl. 1): S2.

280. Yu, H., and M. Lee. 2006. Accessing bioscience images from abstract sentences. *Bioinformatics* 22 (14): e547–e556.

281. Yu, H., F. Liu, and B. Ramesh. 2010. Automatic figure ranking and user interfacing for intelligent figure search. *PLoS ONE* 5 (10): e12983.

282. Zanzoni, A., L. Montecchi-Palazzi, M. Quondam, G. Ausiello, M. Helmer-Citterich, and G. Cesareni. 2002. MINT: A Molecular INTeraction database. *FEBS Letters* 513 (1): 135–140.

283. Zhu, X., and A. B. Goldberg. 2009. *Introduction to Semi-Supervised Learning*. San Rafael, CA: Morgan and Claypool.

284. Zweigenbaum, P., D. Demner-Fushman, H. Yu, and K. B. Cohen. 2007. Frontiers of biomedical text mining: Current progress. *Briefings in Bioinformatics* 8: 358–375.

Index

abbreviations, 12, 15, 19, 55, 69
ABNER system, 73, 76
abstract concepts, 44
abstracts, 3, 5–6, 10–13, 18, 31, 37, 53, 70–71, 88, 91, 94–96, 103–104, 108
accuracy, 24–25, 59, 72, 76, 82–84, 104
active learning, 90
ad hoc information retrieval, 19–20, 33, 35, 94
adjacent labels, 67
adjectives, 14–16, 58–59
adverbial modifiers, 23
adverse drug reactions, 3, 83, 110
algebraic, 42, 44–45
alignment in HMMs, 65
alphabetical characters, 20
alphanumeric characters, 62, 66
Alzheimer's disease, 103
ambiguity, 7, 12, 15, 17–18, 24, 32, 55, 68–69, 104, 108
AMENDA database, 103–104
amino acids, 11, 62
anatomic terms, 27
annotation, 14, 29, 60, 72, 79, 88, 96, 109, 113
annotators, 80, 88
appositives, 17
Arabidopsis, 103
arguments, in relation-extraction rules, 72–73, 75
Arrowsmith system, 107
-ase, as a suffix, 14, 62
assays, 112, 114
assembly, 16
assessment, 94. *See also* empirical evaluation
Association for Computing Machinery, 92
attributes in an ontology, 28. *See also* features
authors, 18, 25
automatically extracted information, 31, 103

background knowledge, 2, 31
base forms, 21
base noun phrase, 23
Bayes. *See* naïve Bayes
Begin/Internal/Other (B/I/O) representation, 60
benchmark, 112. *See also* evaluation
Bernoulli distributions, 43, 50
bigrams, 21–22, 36. *See also* n-grams
binary relations, 69, 74
binary vectors, 49
binary weighting, 39
BIND database, 35
binding, 17–18, 29, 72, 75
bioassays, 112
BioCreative, 92, 94–97, 113
bio-entities, 77, 99, 101
BioGRID database, 35
biological process, 29–30
biomedical connections, 105
biomedical informatics, 2
biomedical knowledge, 10, 26
BioNLP Shared Tasks, 74–75, 92, 97
BLAST. *See* PSI-BLAST system
blood pressure, 16, 36, 38
Boolean combination of terms, 33
Boolean queries, 34–38, 52
Boolean search, 34, 104
Boundaries. *See* sentence boundaries; token boundaries
BRCA, 5–8, 34–35, 55
breast cancer, 5, 55
BRENDA database, 103–104
browsers, 99–100

cancer, 5, 25, 34, 55
canonical dictionaries, 57–58
canonical forms, 57, 95
canonical identifiers, 67–68
capitalization, 12, 20, 57, 66
captions, 11, 18–19. *See also* figures; images
case normalization, 20

catalysis, 13, 16–17, 21–22, 29–30, 54
categorization. *See* text categorization
categorization status value, 46
category labels, 45
causes, in events, 75–76
C. elegans, 103
cellular components, 11, 29, 73, 104, 111.
    See also compartments in a cell;
    subcellular localization
CHAR database, 105
characterization of genes and proteins,
    105, 109–110
characters, 12, 19–20, 22, 59, 62, 71
chemicals, 100–101
chunking, 23–26
citations, 1, 10, 26
classes, 15, 20, 28, 45–47, 50–52, 56, 61–62,
    83–84, 88–89
classification, 34, 45–47, 49, 60, 62–63, 75,
    80, 84, 90–91, 97, 107
classifiers, 45–46, 49–52, 66, 71, 90–91, 97,
    105, 112
clinical notes, 114
clinicians, 3, 5
clustering, 34, 46, 90, 107–109, 112
combination of singular values, 44
combinations of adjacent labels, 67
combinations of terms, 20, 33, 38, 44
combinations of words, 21
commas, 19
common reference corpora, 91
communities, 2
community-wide evaluations, 8, 74
compartments in a cell, 11, 53, 69. *See also*
    cellular components; subcellular
    localization
completeness, 9, 56, 83
complexes, 35, 71
Comprehensive Microbial Resource, 103
conditional independence, 47–48, 52
conditional probabilities, 48, 51, 60, 64, 66
conditional random fields (CRFs), 63, 66–67
confusion matrix, 84
CoNLL-10 shared task, 92
consistency of data, 80, 89, 96
context, for resolving gene and protein
    names, 5, 55, 57, 63, 68, 71
context of entities and terms, 21, 23, 44,
    101–102
controlled evaluation, 79. *See also*
    evaluation
controlled vocabularies, 10, 26, 28–29, 31,
    37
co-occurrence approach, 70
co-occurrences, 70, 101, 104, 106
co-references, 17
corpora, 14, 23, 33, 53, 60, 70, 78, 83, 89,
    91–94, 97–98

cosine-based similarity, 41–42, 68, 111–112
critical assessment of information
    extraction in biology. *See* BioCreative
cross validation, 89–90
curation. *See* database curation
curation bottleneck, 9
curators. *See* database curators
cytosol, 13, 16, 21–22, 54, 74

database curation, 8, 35, 49, 52, 67, 77,
    92–95, 102–105, 108, 110, 112–113
database curators, 4, 9, 35, 37, 48, 52, 54, 80,
    90–93, 102–103, 105, 110–111, 113
Database of Interacting Proteins (DIP), 35
databases, 4, 8–11, 19, 35, 38, 48–50, 52–53,
    67–68, 80, 94, 100, 103–105, 111–112
database schemas, 53
data instances, 51, 81, 83, 90
data integration, 113
datasets, 1, 4, 78–79, 81, 83, 88–90, 95–98
data visualization, 113
decision trees, 46
Defense Advanced Research Projects
    Agency (DARPA), 93
dependencies among words, terms or
    tokens, 15, 22–24, 62–63, 66–67, 73–74, 76.
    See also conditional independence;
    independence
dependency parses, 23–24
determiners, 14, 58–59, 73–74
dictionaries, 21, 56–58, 62, 67, 72, 99–101, 104
digits, 59, 62
disambiguation 23, 69
discourse processing, 15, 18, 93
discourse salience, 69
discovery, 1, 98, 105–108. *See also*
    prediction
discriminative feature selection, 112
diseases and disorders, 1, 4, 17, 27, 34, 36,
    55, 69, 99, 110–111
distributions, 39, 42–44, 47, 50, 60–61, 66,
    83–84, 111
diverse types, 110
document filtering, 34, 52, 68
document frequency, 38, 40
document length, 39–40
document routing, 52
document selection, 39
document vectors, 38
domain experts, 45
domain knowledge, 31, 45, 80, 100
domain-specific, 35, 37, 103
Drosophila melanogaster (fruit fly), 4,
    55–56, 68, 92, 95–96, 103
drug responses, 110
drugs, 3–4, 27, 55, 83, 99, 110–111
drug targets, 111
dynamic programming, 65

EC (Enzyme Commission) number, 104
electronic health records, 113
eliciting meaning, 10, 12, 15, 18, 31–32, 69, 92
Elsevier Grand Challenge, 92, 99
embryonic development, 27–28
emission probabilities, 63–64, 66
empirical evaluation, 2, 8, 60, 77–98
endoplasmic reticulum, 11, 13–14, 16, 53–54
END state in HMMs, 63, 65
end-users, 101, 110, 113
enrichment of terms, 109–110
entity classes, 53–54, 76
entity normalization, 67–69, 95–96
entity types, 56, 60, 62, 67
Entrez database, 67–68
entropy, 60–61, 66
Enzyme Commission number, 104
enzyme information, 103
enzyme names, 14, 104
enzymes, 13–17, 20, 23–24, 54, 103–104
error score, 84
estimation processes, 42–43, 47–52, 64, 83, 89, 104
evaluation. *See* empirical evaluation; assessment
evaluation measures, 80, 93
evaluation metrics, 78
event extraction, 8, 54, 74–76, 93, 95
expectation maximization (EM), 43
expected by chance, 70, 109
experimental evidence, 7, 92, 105
experts, 9, 17, 34, 45, 59, 88–89, 93–94
explicit terms, 42, 44
exponential form, 51
expression data, 77, 113
expression evidence, 92
expression products, 92–93
extraction rules, 60

false negatives, 81–82, 84
false positives, 58, 68, 81–82, 84
fatty acid metabolism, 44, 108
feature dependencies, 66–67
features, 28, 60–62, 66–67, 76, 90, 111–112
feature selection, 112
feature weighting, 66
figures, 11, 18–19, 114. *See also* captions, images
fish oil, and Swanson's method, 105–106
FlyBase, 4, 54, 68, 92, 102
FlyBase curators, 92–93, 105
footnotes, 18
formal evaluation, 77–78
formal grammars, 23
formal languages, 29
free text, 9

FRENDA database, 103–104
F-score, 83–84, 93, 95–97
full parsing, 23
full-text articles, 10–11, 88, 92–94, 96–97, 103, 110
function of genes and proteins, 10–11, 17, 29–30, 37, 94–95, 103, 105, 109, 111

gene clusters and sets, 107–110
gene-disease-drug associations, 110
gene-drug interactions, 110
gene expression, 3–4, 49, 92–93, 107–109, 113
gene identifiers, 68
gene lists, 92
gene mentions, 67–69, 93
gene names, 5–6, 20, 56–57, 67–68
gene nomenclature, 95–96, 107
gene normalization, 96
Gene Ontology (GO), 29–31, 37, 57, 73, 101, 104, 109
gene products, 29, 109
gene regulation, 30–31, 75
gene symbols, 5–6, 16, 55–57, 67–68
generalized dictionaries, 57–58, 62
general-purpose, 23, 35
genetic variation, 3, 77, 110, 113–114
GENIA corpus, 25
genome annotation, 88
genomes, 1, 88, 105, 113
genome sequencing, 1, 113
genome-wide, 107, 110, 114
genomics, 92, 94, 103, 107, 111, 113
genres, 22, 24
global measures of term frequency, 40–41
GO. *See* Gene Ontology
gold standards, 78–80, 88, 96
Google, 33–35, 101
grammatical relationships, 7, 23–24, 36–37
Greek letters, 57, 59, 62

hand coding, 45, 59–61, 69, 71–72, 76
hard classification, 46
headings, 26–27
head words, 73–74
hedging, 7. *See also* speculative statements
held-out data, 89
heteronyms, 16
heuristic rules, 19, 21
hierarchical structure, 26
higher arity relations, 69, 74
high-quality information, 98, 103
high-throughput methods, 1–2, 107, 109–110
HMMs. *See* hidden Markov models
homology, 95–96, 105, 111
hidden Markov models (HMMs), 63–67
hidden variables, 43

homology search, 111
homonymy, 15–16, 38, 44, 55–57, 68
HTML documents, 4, 11, 18, 99
HTML tags, 18
HuGO Gene Name Consortium, 67
human annotators, 88
human-computer interaction, 77
human curators, 90
human experts, 45, 80
human genes, 5, 67, 96, 111
hypergeometric distribution, 70
hyperlinking, 77, 99, 101
hyperplanes, 46–47
hyphens, 15, 19, 59

identifiers, 29, 67, 69, 92, 96, 99–100
iHOP system, 54, 101–102
images, 93, 114. *See also* figures; captions
incompleteness, 6, 57
independence, 47–48, 50–52, 89. *See also*
    conditional independence; dependencies
in-depth analysis, 53, 97
indexing, 10, 26, 28, 31, 35–37, 54, 56, 103
index structures, 36–37
inductive processes, 45
information content, 11
information extraction (IE), 5, 8, 19, 31, 37,
    53–76, 81, 93–94, 112–113
information management, 36
information needs, 3–4, 6, 8, 15, 33, 37, 40,
    42–43, 52, 54, 86, 93–94
information retrieval (IR), 2, 5, 8, 19,
    33–52, 77, 81, 88, 92, 107, 113
input sequences, 63, 65
IntAct database, 35
Integration. *See* data integration
interacting genes/proteins, 1, 6, 34–35, 52,
    54–55, 69–72, 74, 97, 100, 111
interaction curation, 97
Interactive Annotation Task (BioCreative),
    113
interactive systems, 113
interdependent features, 67
interoperability among ontologies, 29
inverse document frequency, 38, 40
inverse document length, 39–40
irrelevance, 6, 33, 37, 49–50, 52, 80–82,
    84–86, 88, 91, 94, 97, 104
IS-A relation, 26, 29–30
IS-PART-OF relation, 26, 29
italics, 18, 57, 71
iterative processes, 39, 43

Jackson Labs, 105
JCVI, 103
joint probability distributions, 63, 66
journal-specific formatting, 18
journals, 10–11, 94, 99

KDD Cup, 92–93
keywords, 35
kinases, 30, 55, 58, 62, 64
knowledge, 1–2, 9–10, 16–18, 25–26, 31–32,
    45, 69, 76, 108, 110

label dependencies, 62–63, 66
labeled as negative, 81
labeled as positive, 81–82
labeled corpora, 22, 32, 61–62
labeled training data, 24, 26, 60–61, 76,
    88–90
labels, 24, 45–46, 56, 60–66, 75, 81–84,
    88–91, 106
label sequences, 65, 67
language models, 42–43
large-scale genomics, 107
large-scale studies, 93, 107, 110–111
latent Dirichlet allocation (LDA), 43
latent semantic analysis (LSA), 44–45, 108
learning-based systems, 59, 62
learning tasks, 46
leave-one-out cross validation, 90
lemmatization, 21
levels of granularity, 18–19, 22, 25
lexical properties, 58, 60, 62, 76
lexical rules, 58
ligands, 17–18
likelihood, 48–51, 65
limited databases, 34
lineages, 28, 104
linear combination, 44
linear functions, 52
LingPipe system, 73, 76
linguistic concepts, 10, 12, 15, 21, 32
linguists, 18, 23–24, 88
lipids, 44, 108–109
lists of names, 56
literature-based discovery, 107
literature mining, 3
local context, 68. *See also* context
local measures, 40
local term frequency, 38
logical axioms, 29
logistic regression, 46, 51–52, 61, 66
lowercase letters, 20, 59, 62

machine learning, 8, 22–24, 26, 32, 43,
    45–46, 51–52, 60–64, 67, 69, 72, 76, 90, 97,
    108–109, 111
manual annotation, 23–24
manual coding, 45, 59, 61, 91
manual curation, 14, 80, 103–104
manual labeling, 62, 88–89, 91
Matthew's correlation coefficient, 112
maximal margin separation, 47
maximum entropy, 60–61, 66
maximum likelihood, 42

mean average precision (MAP), 88, 94
meaning, 9, 12, 14–16, 18, 27, 29, 32, 55, 74,
    88, 111
measure of success, 5
medical informatics, 113
medical information, 114
Medical Subject Headings (MeSH), 10,
    26–29, 31, 37, 91–92, 101, 104
MEDLINE, 10, 26, 37, 91
mentions, 7–8, 54–55, 67–70
merge terms, 59
MeSH descriptors, 27
MeSH qualifiers, 27–28
MeSH subheadings, 27
meta-thesaurus, 37
metrics, 4, 78–80
microarrays, 107
MINT database, 35
MIP database, 19
misspellings, 22
mitochondria, 73–74
mixtures of distributions, 43
model organism databases, 48, 102
model parameters, 48, 50–51, 61, 66–67
modifiers, 23, 74
molecular functions, 29–30
molecular networks, 4, 54, 77, 101–102
morphemes, 14, 32
morphology, 12, 14, 18, 21, 27, 58, 60, 76
motifs, 105
mouse, 3–4, 21, 25, 48–49, 95–96, 103
mouse genes, 52, 95–96
Mouse Genome Informatics database
    (MGI), 4, 48–49, 52, 80, 94, 102, 105
MultiLoc system, 112
multinomial distributions, 42–43, 50
multinomial model, 50
multiple Bernoulli model, 50
multiple sources, 93, 110, 113
mutations, 4, 77, 110

naïve Bayes, 19, 46–52, 71, 97
named entities, 1, 5, 7–8, 11–12, 14, 17, 34,
    53–56, 58–60, 62, 65, 67, 69–70, 74, 76–77,
    81–82, 88, 95, 98–103, 105, 108, 110, 113
named-entity recognition (NER), 7, 19, 21,
    23, 25, 53–58, 60–65, 67–69, 71–73, 75–76,
    99–100, 113
named-entity recognition systems, 56,
    60–61
National Center for Biotechnology
    Information (NCBI), 35
National Institutes of Health (NIH),
    10–11, 35
National Institutes of Standards and
    Technology (NIST), 93
National Library of Medicine (NLM),
    10–11, 26, 28, 35, 37, 91

natural language, 2, 5, 7, 9–10, 12, 14–16, 18,
    31–32, 38, 55, 63, 66, 71–72, 97, 105, 108
natural language processing (NLP), 2, 5, 7,
    10, 12, 15–16, 18, 32, 38, 55, 72, 77, 88, 97,
    105, 108
NCBI taxonomy, 104
negation, 25
negative instances, 81–84, 88, 97
negative regulation, 29, 75
newswire, 24
n-grams, 21–22, 36
nodes in molecular networks, 101–102
noisy data, 79, 90
nomenclature, 95–96, 107
nonalphanumeric characters, 59
normalized weights, 40
noun phrases, 13–14, 17, 23, 74
nouns, 14, 16–17, 22, 59, 61–62, 71–72, 74
nucleotides, 20, 58
number terms, 57
numerals, 59, 62

objective functions, 78–79
OBO Foundry, 29
observation sequences, 66–67
OHSUMED corpus, 91
online journals, 11
online resources, 10
ontological information, 31, 72, 109
ontologies, 7, 10, 26, 28–29, 31–32, 104, 109
Open Biomedical Ontologies (OBO)
    consortium, 29, 31
OpenDMAP, 72–73, 76
organisms, 1, 10, 27, 29, 56, 67, 77, 92,
    99–104, 108
organizational structure, 18
orthographic information, 12, 15, 18, 58, 60,
    62, 66, 76, 100
overall accuracy, 84, 112
over-represented terms, 39–40, 109

paragraphs, 4, 18, 81
parent-child relationships, 26–27, 29
parentheses, 19
Parkinson's disease, 17
parse trees, 7, 16, 23
parsing, 13, 19, 23–26, 97
PART-OF relation, 30
part-of-speech tagging, 13–14, 22–23, 26, 37,
    58–59, 61, 66, 72, 88, 97
parts of speech, 13–14, 21–23, 37, 58, 62,
    72–73, 88, 97
patent records, 4
pathogens, 10, 114
paths through an HMM, 63
pathways, 4–8, 34–35, 110–111
patient records, 4, 114
patients, 83, 114

patterns, 57, 69, 71, 76, 105
PDF, 4, 11, 18
Penn Treebank, 14, 22, 24
Performance. *See* evaluation
periods, 12, 15, 19
Pharmacogenetics Knowledge Base
    (PharmGKB), 110
pharmacogenomics, 103, 110
Pharmspresso system, 110
phenotypes, 110, 114
phosphorylation, 30, 72, 74–75
photographs, 114. *See also* images; figures;
    captions
phrases, 12–17, 19, 21, 23, 32, 36–38, 55,
    72–74, 92, 111
phrase structure, 23
phylogenetic. *See* lineages
physicians, 83, 91, 94
physiology, 27–28
pipelines, 32, 59, 76, 79
platelet aggregability, in Swanson's
    method, 106
PMIDs (PubMed identifiers), 13
polarity, 25
polymorphisms, 111
polysemy, 6, 16, 38, 44, 68
pop-ups, 100
populating databases, 9, 50, 53, 80, 103–104
portals, 9–11
Porter stemmer, 21
positive instances, 81–82, 88, 97
posterior probabilities, 51
potential interactions, 7, 101
precision, 58, 70, 80–88, 95–97, 104
precision at *n*, 86
precision per class, 84
precision-recall curves, 85–87
prediction, 7–8, 22, 58, 62–65, 67, 72, 105,
    111–112
prefixes, 14, 73
prepositional phrase attachment, 16
prepositional phrases, 14, 16
prepositions, 13–14, 16, 59
pre-processing, 71, 101
prior probabilities, 48
probabilistic interpretation, 40, 42
probabilistic latent semantics, 43
probabilistic model, 42–44, 50–51, 63, 108
probabilities, 40, 42–44, 46–52, 63–66
pronouns, 17
pronunciation, 15–16
proper nouns, 12, 17, 22
Protein Data Bank (PDB), 100
protein functions, 95
protein interactions, 4, 8, 19, 35, 52, 70–71,
    96–97
protein localization, 1, 13–14, 53–54, 69–71,
    76, 104, 111–112

protein localization prediction, 112
protein mentions, 95, 97
protein names, 8, 11, 19–20, 54–64, 71–73,
    95–96, 101
protein products, 16, 55, 92
protein-protein interactions, 4, 8, 52, 70–71,
    74, 96–97
protein transport, 69, 72–74
proteomics, 57, 92, 107, 111, 113
providing evidence, 12, 37, 68
pseudo-counts, 50
pseudo-relevance-feedback, 39
PSI-BLAST system, 111
PubChem, 100, 112
PubGene, 101–102
publicly available corpora, 1, 4, 56
publicly available resources, 9, 76, 100, 103
publishers, 11
PubMed, 2, 5–6, 10–11, 31, 33–35, 94–96,
    101–104, 107–108, 112
PubMed Central, 11
punctuation, 12, 19–20

quality of the data, 80
quantifier scope ambiguity, 17
quantitative measurements, 78
queries, 5–6, 8, 10, 20, 27, 33–45, 53, 81, 86,
    88, 90–91, 94, 101, 103, 107
query mechanisms, 38
query terms, 40, 42–43
query vectors, 38

radiology reports, 114
random variables, 47–48, 50, 70
ranked retrieval, 39, 46, 52, 85–86, 88, 94,
    96, 104, 113
rat, 103
Raynaud's syndrome, and Swanson's
    method, 105–106
recall, 58, 80–88, 95–97, 104
recall per class, 84
Reflect system 99–101, 113
regulation, 30–31, 75
regular expressions, 71–72
relation extraction, 7, 53–54, 69–72, 75–76
relation instances, 53–54, 72
relational databases, 9
relations, 5, 8–9, 23–24, 26, 28–31, 53–54,
    65, 69–74, 76, 98
relevance, 33, 35, 39, 46, 50, 52, 81, 83, 91,
    94, 101, 113
relevance feedback, 39
relevance judgment, 91
relevance scoring, 85
relevant documents, 5, 9, 33, 37, 39–40, 49,
    82, 85, 88, 103
repositories, 1, 11–12
representation of documents, 35, 52

representative samples, 79
resources, 4, 7–9, 24, 28, 31–32, 72, 100, 103,
    105, 110, 112–113
retrieval, 20, 33–34, 36, 38–39, 42, 77–78,
    80–83, 85–88, 91, 93–94, 97, 103, 105, 110,
    114
retrieval task, 33–34, 38, 42–43
Reuters collection, 91
RNA, 16, 55
Roman letters, 57, 62
root forms, 20–21
routing, 34
R-precision, 86
rule-based methods, 58–59, 71, 93
rules, 19, 21, 45–46, 48, 51, 58–61, 71–74

Saccharomyces cerevisiae, 12
Saccharomyces Genome Database (SGD),
    102
salient entities, 69
scientific findings, 10
scope of negation, speculation, 25–26
search, 3, 5, 22, 33–36, 38, 70, 92, 101, 103,
    106–107, 110
search engines, 5–6, 33–35, 86, 103
search terms, 38
sections, 18
segmentation of text, 7, 18–19, 71, 111
selection of text, 104
semantic classes and concepts, 31, 70, 72–73
semantics, 20, 29, 42–45, 72, 108, 111
semi-supervised learning, 90
sensitivity, 82, 112
sentence boundaries, 12, 19
sentence fragments, 75
sentence subjects, 16, 23
sequences, biological, 1, 11, 20, 77, 100–101,
    105, 107–108, 111–113
sequences of characters, 12, 20, 36
sequences of states, 63–64, 66
sequences of tokens, 12, 23, 54, 58, 60,
    63–66, 72
sequential models, 63, 66
shape features, 62, 66
shared evaluation tasks, 74–75, 91–98
SherLoc system, 112
significance. *See* statistically significant
    differences
silent states in HMMs, 63
similar documents, 10, 20–21, 33–45
similarity measures, 33–45, 112
similarity queries, 33–45, 52
singular value decomposition, 44
singular values, 44–45
skewed distributions, 83–84
slots in ontologies, 73–74
small datasets, 90
small molecules, 99–101

SMART database, 100
soft classification, 46
source models, 43
species, 20, 55, 68, 95, 105
species name recognition, 68
species names, 12, 68, 104
specificity, 82, 84, 112
speculative statements, 25–26
spelling, 12, 15–16
standard tokenization, 91
START state in HMMs, 63
states in HMMs and CRFs, 63–68
statistically significant differences, 79, 96
stemming, 16, 21, 31, 36, 38, 97, 108
STITCH database, 100
stochastic. *See* probabilistic model;
    probabilities
stop-words, 38
string matching, 56
structured databases, 8–10
structured representations, 1, 8–9, 11, 32,
    53, 74
stylistic features, 18
subcellular localization, 53–54, 69, 100, 104,
    111–112
subscripts, 18
subsections, 18
substrings, 31, 62, 75
suffixes, 14, 62, 73
summarization, 37
superscripts, 18
supervised machine learning, 22, 32, 46, 90,
    107
support vector machines (SVM), 46–47, 52,
    97, 112
surrogate representation, 107
Swanson's method, 105–107
Swiss-Prot database, 11, 111–112
synonymy, 6–7, 16, 26–27, 31, 38, 44, 68, 92,
    98, 100–101, 104
syntactic ambiguity, 16
syntactic analyses, 23, 25, 70, 74
syntactic information, 60, 72, 74
syntactic roles, 14, 59
syntax, 9, 18, 29, 72

tagging, 13, 18, 22–23, 37, 58, 82–83, 88, 91,
    97
taxonomies, 11, 26, 28, 31, 68
teams, 94, 96–97, 104
term combinations, 44
term frequencies, 38, 40
term occurrences, 40, 42, 49–50
term weights, 38, 47
terminology, 52–53, 80
terms, 5–6, 10–11, 20–22, 26–31, 33–34,
    36–45, 47–52, 55, 57–59, 62, 68, 81–82,
    101, 104, 106–109, 111

test sets, 89, 93
text-based features, 111–112
text categorization, 8, 33–35, 37, 45–46, 50, 52, 84–85, 88–92, 94, 103, 107
text classification, 45, 50, 90
text passages, 52
Textpresso system, 103–104, 110
text representation, 1, 9, 32, 36, 107
Text Retrieval Conference (TREC), 88, 93–94
text sources, 3–4, 7–8, 10, 53, 69, 72, 112
text visualization, 113
textual context, 71
TF-IDF, 40–41
thematic analysis, 43, 50, 107–110
themes in theme models, 43, 50, 75–76, 108–109
thesauri, 26, 28, 31
three-dimensional structures, 100
tissues, 29, 104
titles, 18, 31, 104
token boundaries, 19, 56
tokenization, 19–20, 91
tokens, 12, 18–20, 22–23, 36, 50, 56–57, 59–66, 71–72, 74, 111
topic models, 42–43, 50, 107–109
top-ranking documents, 85
traceback in HMMs, 65
trade-off, precision and recall, 82–83
training corpora, 32, 62
training data, 26, 45, 48–52, 60–61, 64, 67, 89–90, 93, 96–97
training examples, 46–47, 60, 76
transcription, 55, 69, 92–93
transformed vectors, 45
transition probabilities, 63
transitions in HMMs, 64
transitive links, 105
translocation, 73–74
TREC Genomics track, 3, 92–94
treebanks, 24–25
trees, 13–15, 27–28, 46
triggers, 75–76
true negatives, 81–83
true positives, 81–83, 102
two-dimensional structures, 100
types and tokens, 20
typography, 15, 56–57, 68

ubiquitin, 11, 13, 15, 17, 23–24, 54
unambiguous parses, 11
Unified Medical Language System (UMLS), 28, 37
unigrams. *See* n-grams
UniProt, 37, 67, 103. *See also* Swiss-Prot database
UniProtKB, 11
unique identifiers, 29

unlabeled instances, 88, 90
unstructured representations, 53
unstructured text, 9, 11
unsupervised categorization, 46, 90, 107
uppercase letters, 58–59, 62
user information needs, 4, 6, 37, 40, 42–43, 52, 113
user satisfaction, 77

vector model, 38–46
vector representations, 38–47, 52, 112
verb lists, 72
verb phrases, 13–14, 23
verbs, 14, 16, 20, 23, 71–72, 101
visual information, 110, 113–114
Viterbi algorithm, 65, 67
vocabularies, 9, 11, 26–28, 38, 42, 50, 52
vocabulary of a database, 38

web, 1, 10, 34, 99–101
websites, 10–11, 34, 102
weighting schemes, 38–41
weight vectors, 39, 43–44, 47
wild type, 92
word dependencies, 15, 22
words. *See* terms
WormBase, 103

yeast, 11–12, 95–96, 108
yeast genes, 16, 95

zebrafish, 103